A Young Person's Guide to Military Service

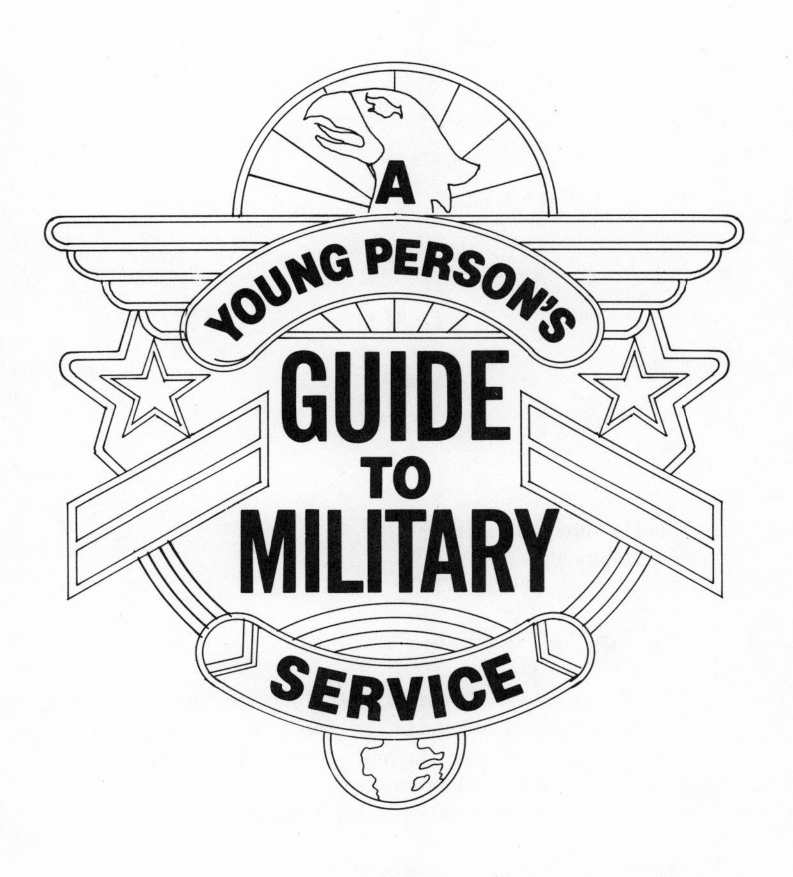

Jeff Bradley

Foreword by U.S. Sen. Edward M. Kennedy

THE HARVARD COMMON PRESS
Harvard and Boston, Massachusetts

To Marta

The Harvard Common Press
535 Albany Street
Boston, Massachusetts 02118

Printed in the United States of America

Library of Congress Cataloging-in-Publication Data

Bradley, Jeff, 1952-
A young person's guide to military service.

Bibliography: p.
Includes index.
1. United States—Armed Forces—Vocational guidance. I. Title.
UB323.B7 1987 355'.0023'73 86-33719
ISBN 0-916782-82-4
ISBN 0-916782-83-2 (pbk.)

Cover design by Peter Good

10 9 8 7 6 5 4 3 2

Contents

Foreword

Life in the American military is very different today from what it was when my three brothers served in the Navy in the 1940s, or when I was P.F.C. Kennedy in the 1950s.

One of the biggest changes has been the active recruitment of women. In 1976, women were first admitted to the service academies, and since then, their numbers have grown rapidly—in the officer corps as well as among the ranks of the enlisted.

Another change has been the elimination of the draft and the development of a highly qualified, highly trained all-volunteer force. The change has been a success, because the service has worked hard to attract outstanding young men and women, to make military pay competitive with the private sector, and to improve the quality of life for everyone in uniform.

National advertising campaigns now advocate careers in military service: the Navy tells us, "It's not just a job, it's an adventure"; the Army urges, "Be all that you can be"; the Air Force says, "Aim high"; and the Marine Corps announces, "We're looking for a few good men."

A third profound change comes from increased reliance by the armed forces on sophisticated technology. Smarter and more complex machines require brighter and better-trained men and women to run them. Computers will never replace the soldier in the field with helmet, rucksack, and carbine, but the technological revolution is transforming the military and will continue to do so in the future.

Since 1983, I have had the privilege of serving on the Senate Armed Services Committee, which has responsibility for guaranteeing that our defenses are strong and that our forces are well trained and well equipped. In the past few years, I have made a point of seeking out men and women in uniform to learn directly about life in the armed forces.

I have traveled to U.S. military installations throughout the world and talked with American servicemen and -women performing difficult missions under challenging conditions. I met with G.I.'s in the middle of winter at the Fulda Gap in Germany and talked with soldiers on duty in the Sinai Peninsula as members of the Multinational Force and Observer Group. I met with sailors and pilots serving on a nuclear-powered aircraft carrier in the Atlantic and sailed with the crew of an attack submarine in the Mediterranean. I flew with an air force pilot in an F-16 and walked the flight line with the best fighter pilots in the world at Nellis Air Force Base. I watched the Marines land and take the beach at Camp Pendleton, and I flew with a Top Gun Marine pilot in an F-18 at Miramar Naval Air Station in California.

The quality of Americans in uniform has never been better, and their spirit has never been higher. They are dedicated, committed, and courageous.

At the same time, I also believe that we would be a lesser nation in a more dangerous world if we did not recognize that it is as important for America to have a Peace Corps as it is to have a Marine Corps. National security is more than merely safety from external threats. It is also national confidence and pride in what America stands for—a willingness to sacrifice for others, to serve our country, to care about those in need, and to be active participants, not passive spectators, in America's commitment to freedom, justice, and opportunity.

There are many reasons to join the military—and other reasons not to. Jeff Bradley's book is the best written and most balanced guide to military service available today. I urge any young man or woman who is thinking about spending time in the military to spend a little time reading this book first.

It's one of the most important decisions you'll ever make, and this is the book that will give you the information to make it wisely.

SEN. EDWARD M. KENNEDY

Preface

This year over three hundred thousand young people will enter the armed forces. Many will join to take advantage of vocational training or earn money for college. Some are looking for adventure, and others want to travel. A good many people will join to get away from something—home, a dead-end job, or a failed romance.

I wrote this book for two reasons. The first is to present a balanced view of life in uniform. When young people seek information about the armed forces, virtually all they find comes from the Pentagon. The brochures and pamphlets have slick paper, exciting color photographs, and a persuasive message. This shouldn't be surprising, for most of the recruiting materials are prepared by the same advertising agencies that so successfully sell shampoo and microwave popcorn.

With all this razzle-dazzle it is difficult to get an even-handed picture of the military. Some people say it can be the best thing for a young person, and the relentless T.V. commercials certainly make it look wonderful. For many people in uniform, the military is a good experience. There is, however, another side to the service that seldom gets mentioned. Military life is vastly different from the civilian world. Some people never make the adjustment and are profoundly unhappy the whole time they are on active duty. And amid all the talk about jobs and money for college, one feature of the military rarely comes into focus: combat.

The groups that strive to present an alternative view of military service have ant-sized budgets compared to the Pentagon's. In addition, many of these organizations are closely tied to political or religious views outside the mainstream of this country. This further reduces their chance of being heard.

Drawing from several points of view, I have tried to present an objective picture of life in uniform. My sources of information have ranged from career officers in the Pentagon to professional peace activists. This book does not try to talk anyone in or out of joining; no evenhanded book could do that. The message of this volume can be expressed in one sentence: Eight years is a long time; make sure you know what you are getting into.

My second reason for writing the book was to present, in one place, as much information about various military options as possible. When I wrote the first edition of this book, I assembled a stack of military brochures and pamphlets over a foot high. Their contents were often vague, sometimes confusing, and occasionally virtually useless.

In this second edition, I have completely updated the existing information and added three new features. Now each of the chapters on the various branches contains a list of places where more than 100 men or women are stationed. Each of these chapters, too, includes a complete listing of entry-level jobs, and an appendix in the back of the book lists all the colleges and universities that currently offer ROTC programs.

The experiences and comments of hundreds of people have shaped this book. It is impossible to name the young people, active-duty personnel, reservists, and veterans who have taken the time to talk with me. The following are people in various organizations who gave me special help with this project, whether by providing information, answering questions, or checking chapters for accuracy.

From the Army, Col. Paul Knox and Col. Alexander Woods, Jr. From the Navy, Anna Urband and Lt. Com. Bill Clyde. From the Air Force, Maj. Edward Stanford. From the Marines, Maj. Anthony Rothfork, Sgt. Maj. Clarence McClenahan, and Maj. James Vance. From the Coast Guard, Lt. Com. Phil Matyas and Lt. M. J. Smith. From the National Guard, Kathleen Jewell and Col. Donald Consolmagno. Anita Lancaster of the Department of Defense looked over the manuscript.

Robert A. Seeley of the Central Committee for Conscientious Objectors was kind enough to look over portions of the book, and Greg Sommer of the Midwest Committee for Military Counseling provided some information and good advice.

Larry Anderson came up with the idea for this book, Kathleen Cushman recruited me for the Harvard Common Press, and Linda Ziedrich edited the book. John Marius, Com. Sgt. Maj. Ray Roy, and Sgt. Allison Staton, Maj. Kenneth Carr posed for my camera; and Neal Bierbaum once again shared his thoughts on the military.

Finally, thanks must go to those Harvard freshmen in my writing classes who chose to visit Boston military recruiters and write about their experiences. In grading their papers I learned a lot about the recruiting process that I would otherwise have never known. The students came back with a new perspective on the military, and the unsuspecting recruiters got a chance to see some high-powered potential troops. The recruiters will probably never forget the young man who sat down, took a forty-five minute test in ten minutes, and rolled up a perfect score. They are probably still after him to join.

1. Thinking about the Military

Photo by author

FOR MOST YOUNG PEOPLE, life seems to be a bowl of choices. You are faced with decisions on what classes to take, what schools to attend, if and when to get married, and what sort of work to choose. Many of these decisions are interrelated, and none of them are easy.

If you ever listen to the radio, watch television, read magazines, or look at billboards, you find yourself hit with another set of choices: these involve military service. The Army, Navy, Air Force, Marine Corps, and Coast Guard are spending $219.2 million in 1987 to get you to think about joining up.

You see the ads everywhere. Some promise jobs, others stress dollars for education, and some push "the adventure of a lifetime." After a while it begins to sound like you can join up and become a fighter pilot, get paid to go to college, live in an exotic place, learn a trade that promises a lot of money, and have a wonderful time doing it all.

Often the ads focus on a particular problem of yours. If you have been looking at a college catalog and trying to figure out where you will get the money to pay for your schooling, your ears perk up when they talk about cash for education. If you're stuck in a low paying job and have no marketable trade, you listen when they talk about job skills. And if you're tired of home and sick of school and bored with your town, you dream of going to places like Europe, Hawaii, or the Far East.

You begin to think about getting into the military.

If you tell anyone what you're thinking about or ask them for advice, you'll probably get an earful. Whether they've been in uniform or not, almost everyone has an opinion about the armed forces. Many people think the military is somehow good for you. Others express shock that you want to have anything to do with the armed forces, whom they associate only with death and destruction. And your friends may laugh at the thought of you doing pushups while wearing a uniform and short hair.

They've all got something to say and a clear idea of what you should do with the next few years of your life. But it's *your* decision, and there are a lot of things to consider as you make it. Military service can be a very satisfying experience with some distinct benefits. Joining up can be the first step in solving some of the problems in your life. But it can also be an enormous source of frustration and unhappiness. Once you are in uniform the military has control over you—more control than you may have bargained for.

The Advantages of Military Service

Joining one of the branches of the armed forces may very well be a good way of straightening out your life. On a very basic level the military can remove some of the hassles of your day-to-day existence. When you join up they give you new clothes, a clean place to live, and plenty of food that you don't have to cook or pay for. If you get sick or need to see a dentist, it doesn't cost you anything. No one ever claims that military life is luxurious—but neither is living in a crummy apartment and subsisting on a diet of junk food.

While your basic needs are taken care of, you can learn a useful skill. The military has hundreds of jobs that need people to fill them. Virtually everything in civilian life—banks, hospitals, schools, stores—has a counterpart in the armed forces, and the range of positions is wide. The trucks, typewriters, tanks, computers, and ships all require maintenance and repair, and someone will teach you how to do it. Best of all, unlike most civilian jobs, occupational specialties in the military require no previous experience. Several jobs are worth a bonus of thousands of dollars just for completing the training. Try to find a deal like that on the outside.

For those with sights on higher education, military service can cut college costs considerably or eliminate them altogether. You can take courses while in uniform, and once you get out you become eligible for veterans' education benefits—even if you've never been in combat. If you've already been to college and want to go to graduate school or even medical school, it's possible to have all your costs taken care of and receive a salary as well.

Even more important than jobs or education, the military

can be a place to make fundamental changes in your life. For the young man who has dropped out of school and finds himself on a downward path, the military can provide a place to turn things around. For the young woman who is stuck in a low-skilled, low-paying job with no chance for promotion, the armed forces can give her a chance to move up in the world. And for the person who wants to get away from home and find out what he or she is really capable of, the military can offer an excellent place to begin.

Providing such advantages used to be called "making a man out of you." Those who used this phrase meant that during a hitch in uniform people sometimes changed for the better. They came back more disciplined, able to accept criticism without flying off the handle, and willing to take responsibility for their actions. And, most of all, they came back with a firm grasp of what they wanted to do—no longer the victim of pressures from family or friends. This feature of military service still exists, and for some people can be the best benefit of all.

Finally, for some individuals there is great satisfaction in serving their country, defending it from attack, and protecting Americans and their interests wherever they may be. You cannot put a dollar value on this good feeling.

The Disadvantages of Military Service

The good things that come from the service are not there because the government wants to be generous. You are expected to give in return. When you take the oath of allegiance, you surrender control of your life. The men and women above you will decide and act upon what is best for the military, not necessarily what is best for you. The decisions they make can cause you serious injury or even death. One has only to see the rows of markers at a military cemetery to grasp this.

The military has no equivalent in civilian life. Nobody makes a lot of money in uniform, and few people who wear one win any fame. There is good reason the armed forces are referred to as the *service*. Nonetheless, the people in the military are charged with perhaps the most serious of tasks: defending the country against those who would harm it. This number-one priority sometimes gets lost in all the talk about jobs and education.

Some aspects of military life you'll rarely see in the ads or brochures. Engaging in combat, even on a small scale, is a nasty business. People in uniform get injured—sometimes disabled for life—and killed. Planes get shot down, ships get sunk, and hostile forces shoot back. Even people not in uniform can get hit, and sometimes innocent people are killed. This happened during World War II, during the Vietnam War, and as recently as the attack on Libya, in which dozens of innocent Libyans died in the night. If you are involved in a combat operation, no one will ask your opinion on the rightness or wrongness of it. You will be ordered to carry out actions to the best of your abilities, no matter what you think.

The tight control of your life that is necessary in wartime is present all the time in the military. You lose certain rights that civilians take for granted; one of the first to go is free speech. Your commanders can order you to perform any task, no matter what you trained for or would prefer to do. They can move you from base to base and even out of the country. They can control your promotions, vacations, appearance, and even the way you talk and act.

And if you don't like it, you can't just quit.

How to Use This Book

Whether or not to join the military is one of the more important decisions you will ever make. You need to consider carefully what you want out of life and whether these goals can be attained in the military or somewhere else. You should weigh the advantages of the services with the drawbacks that come from putting on a uniform. You need to learn as much about your options as you can.

This book can help you accumulate knowledge about military life in general and individual branches in particular. It combines nuts-and-bolts facts with questions to stimulate your thinking. The first five chapters outline various options you have—officer or enlisted, active duty or reserves, and things like that. Most of these options apply to the entire service, so after digesting this material you will be ready to look at the separate branches in greater detail. You can then talk more confidently to recruiters or seek the advice of military counselors. Places to find counselors are listed at the end of Chapter 12.

This book will not make your decision for you; no one book or person can do that. It won't even make a recommendation. Except one: before you sign papers that may commit you to eight years of military service, take enough time to learn all you can so you can make the best decision possible.

2. Questions to Ask Yourself

Official U.S. Navy photograph

After thinking about it and talking with a few people, you may begin to think the benefits of military service look pretty appealing. So you walk past a recruiter's booth in a shopping mall or in your high school and pick up some of the brochures.

Before your eyes is spread a feast of educational programs, job training packages, and enlistment options. There are color photographs showing people flying helicopters, operating computers, or scuba diving. You may see a young man working on an engine as big as a room, or a young woman making some technical adjustments on a sleek jet fighter. Or you may simply see people having fun in an exciting place.

These glossy brochures answer all sorts of basic questions about how many jobs are offered, where you can travel, and how long you have to stay in. There are some larger questions, however, that are seldom addressed in this material. An examination of these matters, many of which aren't always apparent to someone outside of the armed forces, can clarify your values and give you some things to think about before joining up.

Here are twelve questions to guide your thinking. They are grouped in four areas: your goals, the military concept, the military system, and your personality.

Your Goals

When we look into the future, we rarely see far. In signing up for the military, however, you are affecting eight years of your life. It's worth taking some time to assess why you are considering the armed forces.

Where do I see myself after my tour of duty? What preoccupies most new recruits is basic training. When this is over they focus on what happens next, and so on. Think for a while about longer-term goals. At the end of your active duty, do you envision yourself with a college degree, a family, and a satisfying job? Are you looking for a well-paying skill that will enable you to live the kind of life you want to live? Do you want money to pay for an education? Or are you looking for some good times in interesting places?

Once you get a sense of where you want to be in eight years, you'll have a better idea of what to look for in the military. If, for example, you are looking to earn a college degree as soon as possible, you will then be able to scan the brochures and talk to recruiters with more authority. Best of all, you will be less likely to get talked into something that doesn't fit your needs.

Can I get what I want outside the military? The armed forces are a frequent choice because they advertise a lot, they are easy to enter, and many people approve of them. Some goals don't require a uniform, however. You might be considering joining the service just to get away from home and your parents. But you could just as well get away by going to a distant city with a friend and finding a job there. This could be difficult, and there would be no guarantee that you'd succeed, but you have no promise that you'll attain your goals in uniform either.

If an education is what you are after, you might check with a good guidance counselor or someone else who can give you competent advice. Most colleges and universities have financial aid or work-study programs that can hold the academic costs down. The reason you may not have heard about these programs is that together they don't have $200 million to spend on ads. Harvard, for example, though one of the more expensive colleges in the country, offers a complete financial aid package to every student who needs it.

Am I thinking of joining just to please someone else? Your father may be after you to join up and "straighten yourself out." You may have a friend who is in the Army and who is trying to get you in. You may want to please a high school teacher or even a recruiter. You may want a uniform to impress members of the opposite sex.

These are probably the worst reasons to join the service. *You* will be the one in basic training doing pushups. *You* will be the one who might be stationed in Greenland scraping ice off vehicles. And *you* will be the one who is obligated for eight years. Therefore, you should join the military only if *you* want to.

The Military Concept

Every war that the United States has fought has been opposed by one group of citizens or another. Some objected to the use of force at all, while others were against a particular conflict. These feelings reached a peak during the Vietnam years, when thousands of people took to the streets and songs of protest reached the Top Forty. Young men began to seriously question whether they could in good conscience take part in the armed forces. It's a question everyone considering military service must think about.

To help clarify your own ideas, you might write to one or more of the antimilitary groups listed in Chapter 12. Ask them to send you some literature. What you will get back in the mail won't be glossy and expensive like the materials you'll get at the recruiting offices, but the point of view expressed may stimulate your thinking. As you look at both military and antimilitary brochures, of course, you should keep in mind that both sides are biased.

Do I accept the use of force as a tool of national policy? Most people believe that the United States should use force to protect its citizens and defend the country. When Adolf Hitler talked about taking over the world, almost everyone saw the need to stop him.

Even when such a threat is less obvious, some policy makers contend we are obligated to oppose it. The war in Vietnam was fought to halt the spread of communism in another country. Recent events in Central America have once more brought up the question whether a small nation's choice of political system can possibly threaten the much more powerful United States. As in Vietnam, the issue is not clear-cut.

Some strategists go further: they contend that the United

States is justified in attacking groups or countries suspected of harboring terrorists. There is even less agreement here.

As administrations and public opinion change, the emphasis on the use of force varies. A person might join the armed forces when things are pretty quiet, and an election or two later find the climate has been altered entirely. Yet that person must carry out orders from Washington without question, regardless of how he or she feels about them.

Can I take part in something I do not support? Obviously if you are against bearing arms you will not spend much time hanging around the recruiter's office. In all the talk about jobs and education, however, it's easy to lose sight of some of the things the military may be called upon to do.

How do you feel about nuclear weapons? Would you be comfortable working in a missile silo or servicing a bomber that carries nuclear weapons? You might be willing to defend the United States, but how would you feel about fighting against insurgents in Central America? If you served in a National Guard unit, would you have any objection to halting a riot in an American city? Or protecting strikebreakers in a labor dispute?

Outside the military you are free to support whatever cause you want. You can act on your principles by opposing something you do not like, whether your opposition takes the form of discussion or participation in a demonstration.

Inside the military, however, dissent has no place. You are expected to carry out orders and fight where told regardless of your convictions. Are you willing to do this?

The Military System

Steeped in tradition and usually run by the most senior officers, the military often seems a couple of steps behind the society as a whole. Changes do take place, but they are slow in coming. It's worthwhile to examine the way things operate before putting on the uniform.

Can I operate in a large organization? If you are like most eighteen-year-olds, the largest organization you have been a part of is your high school. You may not have been personally acquainted with the principal or headmaster, but at least you knew what he or she looked like.

The military is a vast, sluggish bureaucracy where the chain of command is followed as if it were a lifeline. Sometimes the simplest of requests must pass through a dozen unseen and unknown people, any of whom can turn it down or lay it aside.

It takes special ambition and drive to succeed in this mass of people and paperwork. Some adapt quickly, learning the official and unofficial rules and using them advantageously. Others become frustrated at the system and procedures and seem to be in constant battle with those above them. How would you fit into such an organization?

Can I adapt to the military way of doing things? The armed forces stress uniformity, and they have particular ways of doing some of the most common things. Often the greatest frustration can come from these small matters.

In the Army's Officer Candidate School men and women are given a diagram showing them exactly where everything on their desk is supposed to go. Included is a paragraph on how handkerchiefs are supposed to be folded inside drawers.

One male Air Force officer was told it was unbecoming for him to be seen on base carrying a baby. This meant that his wife, no matter how tired or uncomfortable she was, had to hold their child as the two walked from place to place.

These are small matters, to be sure, and probably seem worst at first encounter, but they are indicative of a rigidity that can prove highly annoying to some people.

Can I adapt to moving around a lot? Seeing new places is one of the attractions of the military, as anyone who has been stationed in Hawaii or Europe can testify. After a while, however, the thrill of picking up and moving is replaced by the hassle of packing possessions, leaving newfound friends, and starting all over again.

The situation is worse if you have a family. It is virtually impossible for the spouse of a military person to have a meaningful job. And it's hard on children to leave their friends and change schools every year.

For the single person, travel can be exciting, but Christmas or another important holiday is no fun when you're sitting in Turkey while the rest of your family is at home eating one.

Your Personality

Some people seem to thrive on military life, while others hate every minute of it. Deciding whether or not to join involves taking a look at yourself and predicting how the partnership will work out.

Do I take orders well? This seems obvious—if you can't take orders you had best start near the top or stay out altogether.

Following instructions is a vital part of most jobs, yet in the civilian world there is often a chance for discussion and compromise. This is not always the case in the military, particularly during a person's first years there. You may find yourself following orders without a chance to say anything at all, even if you know a way of performing the task more efficiently.

Whether as an officer or as an enlistee, you may be ordered to perform tasks that you don't want to do or that have no apparent usefulness. These orders may come from someone who is younger than you, who has less experience or education, or who you feel is unfit to give orders. Whatever the case, you have to do as you are told; no arguments are allowed.

Can I overlook offensive remarks and attitudes? From the academies on down to boot camp, newcomers to the service face a lot of harassment. Whether from a drill instructor shouting in your face or some lout in the barracks, you are likely to hear derogatory comments about your appearance, sex, race, and any aspect of yourself that draws attention.

Women in the armed forces, particularly enlistees, are likely to have sexual remarks directed at them. These can range from off-color jokes in your presence to outright invitations to engage in sex.

Blacks and members of other minority groups may be

subject to racist remarks. For someone who has never let another person get away with such behavior, this can be very difficult to take.

The military is sensitive to charges of sexual and racial harassment, and on an institutional basis it works very hard to prevent it. It cannot, however, prevent individuals, particularly those who hold rank over others, from making sexist or racist comments.

Do I have unrealistic expectations of the armed forces? Sometimes you can get caught up in the power and glory of the ads and get a false impression that military life is going to solve your every problem. Anyone coming in with this attitude is in for a letdown. Many people have made significant changes in their lives while in uniform, and much of the credit can be given to the military experience. But the change has to come from within.

To get a balanced picture of what to expect, you should not only consider what the military offers but what you bring with you. If, for example, you have been fired from your last three jobs, you might do well to consider why this has happened and what you can do to remedy the situation.

As one former Navy chaplain explained, "Some people can go into the military and shape up their lives. But the military will also give you an opportunity to fall flat on your face."

The appropriate question then is: What do I expect from military service, and what can I do to realize my expectations?

What if things don't go as planned? Often people enter the military with a specific goal in mind. A young man may go to the Air Force Academy intending to become a fighter pilot. When he finds he cannot pass aviation training because of severe nausea, he may become despondent. Or a young woman enters the Army wanting to become an air traffic controller. She cannot pass a security clearance and is put in a job that involves mostly typing—just the sort of thing she had tried to get away from in civilian life.

Sometimes things do not go as you planned. This may be due to a failing on your part or a situation beyond your control. Although an effort is made to accommodate the wishes of those in uniform, the military is not geared to please the individual. Someone has to guard empty barracks at Fort Campbell, Kentucky; someone has to run the radar station near the Arctic Circle; and someone has to work in the hot engine room of a ship.

You should keep in mind the possibility that nothing will turn out as planned, and that your two or four years on active duty will seem like eight or ten. No one likes this, neither the military authorities nor the individuals affected, but it can and does happen.

If you put all of the situations discussed in this chapter together, the military can sound like one awful place. Lots of people, however, come into the service, make it through boot camp with a minimum of hassles, complete the training they wanted, and happily work in the field of their choice. Others find they have to change their plans along the way, and for some the whole thing is plain hell.

In talking with veterans of military service, a phrase one hears again and again is "the proper attitude." Ex-Marine Bob Abernathy puts it best. "Attitude is the whole thing in the military. *You* make what you are in the military."

3. Where Will You Fit In?

Official U.S. Navy photograph

ONCE YOU DECIDE you can operate in the military, you still need to focus on what you want out of it. Only then should you start selecting among the various branches and the options within them.

WHAT DO YOU WANT OUT OF THE SERVICE?

In some respects all the various branches of the armed forces are the same. All of them will get you up at the crack of dawn during basic training, and pay and promotion schedules are roughly similar. Every branch offers job training, and benefits come to all veterans who have been given good discharges.

Despite these similarities there are a great many differences. In the Air Force, for example, the officers are involved in most of the combat while the enlisted people stay on the ground. The Navy offers you a chance to travel all over the world, whereas in the Coast Guard you may find yourself exchanging shots with drug smugglers off the coast of Florida. A soldier in the Army may never experience any excitement, whereas a National Guard member may assist during floods or other natural disasters.

Making these decisions forces you to examine why you are getting into the military. For some people this is easy. "From the time I was in the Scouts I wanted to be a pilot," said one Air Force captain. "For me there was no other way to go." Others come from a tradition of serving in a particular branch, and perhaps want to follow in their father's footsteps. For someone coming from a background like this, there's no "which" to decide. For others, an examination of what the military offers is in order.

The Adventure of a Lifetime

For a lot of people the military is just what the ads depict: jets zooming into action off an aircraft carrier, tanks rumbling across a stream, or an elite commando group making its way through a jungle. These people thrive on the excitement the military offers. Patriotism and a desire to serve their country may be a big motivation and acquiring job skills or going to school is secondary.

"I was right out of high school," said one Army officer. "Before long I was a company commander in Vietnam. There I was, a guy in a hole with a radio, orchestrating artillery and air strikes—telling the Navy and the Air Force what to do. The Vietnam experience was the most real thing in my life."

If military hardware such as tanks, combat aircraft, and guns of all sizes appeal to you, you should head straight for the more action-oriented divisions of the service. Being a machine gunner in the Marines may be your idea of what a real military career is all about. If you want to save money for your education, taking a combat job will help by earning you the largest bonus.

In addition to large numbers of infantrymen, tank crews, and artillery specialists, in the combat forces you will find the elite units such as the Army Rangers or the Navy Seals, distinguished by their rigorous physical training, discipline, and combat abilities. These units command a vast amount of respect in the service and in the civilian world. "He was a Green Beret" is a phrase that will follow a man for the rest of his life.

These specialized units are highly selective, and to get in you have to have outstanding physical and mental qualities. One drawback to being in an elite unit is the constant training to maintain a readiness that is seldom called upon. A unit of Army Rangers, for example, found themselves in the blistering hot Sinai desert keeping the peace between Egypt and Israel. It was an important mission, but a mortifyingly boring one. Keep in mind that the skills men work so hard to master here are not always directly applicable to civilian life.

Paying for Education

Not everyone who considers military service is gung-ho about the action and the weaponry. Perhaps the fringe benefits are your primary concern. You approach the military with a business proposition: I'll give you my time in return for an education.

Although Chapter 15 is entirely devoted to educational opportunities, here are the main considerations. More educational options are available to officers, who can attend classes before, during, or after active duty. Enlisted personnel have several choices, however, including one that will pay 90 percent of tuition costs for classes taken during off-hours.

If you want to go to college after you get out of the service, and are intent on amassing money in the meantime, then you should look for a military specialty that offers a high bonus. These jobs usually fall into one of two groups: those that require a long period of training, and those that are dangerous or most likely to involve combat. You take your pick.

If you want to take classes while in uniform, you should look for a branch that offers you a chance to live close to colleges and universities. Correspondence courses are offered in every branch, but the best education takes place in a classroom presided over by a professional teacher. This rules out submarine duty, and you might try to avoid getting placed at a radar station somewhere in the frozen north.

Getting Job Skills

After working as a hamburger flipper or car wash attendant, many young people join the military to gain marketable job skills. Every branch offers a wide array of positions that teach technical and "people" skills, such as telephone repairing and hospital management. And all of the branches offer training in computer programming and other "high tech" fields.

For more specialized skills, however, each branch of the service varies. If you want to work with diesel engines, it makes sense to head for the Army, with its vast numbers of trucks and tanks. If you seek maritime experience, then the Coast Guard or the Navy will be your ticket. When you spread out the brochures from all of the branches it looks as if you have hundreds of jobs to choose from, all with pictures of men and women who seem to be enjoying themselves immensely.

In practice, however, the list you will choose from will be much shorter. Your abilities, your sex, the needs of the service, and the length of time you plan to stay will all reduce the hundreds of possibilities to a more manageable number. And no matter which branch of the service you approach, you'll go through the same procedure to determine the job for you. (This procedure is described in Chapter 4.)

The Thing the Ads Never Mention

The ads and posters never show anyone getting wounded or killed. This is the aspect of military service that no one likes to think about, yet it should guide your choice of which branch to join.

If hostilities break out—even a skirmish—the forces that take part are likely to suffer casualties. Even though the American planes had the most sophisticated electronic defenses in the world, two U.S. fighter-bombers did not make it back from a brief raid on Libya. Three years earlier, in Beirut, Lebanon, 241 Marines died in a terrorist attack, and they weren't in combat at the time.

Certain air and ground forces, such as the Marines, are traditionally the first ones to enter a fight. If you are in such an outfit, your chances of coming home wounded—or not coming home at all—are increased.

Few people look forward to armed conflict, but the military forces are constantly training and preparing for it. If you have chosen a combat job, when your unit gets into a fight, so will you. It's worth considering where you want to be when the shooting breaks out.

ACTIVE OR RESERVES?

All of the branches of the service are broken down into two main components—active and reserve. Members of both groups receive the same training, use the same equipment, and are paid at the same rate.

The big difference is how much time they spend in uniform. Active duty is just that—you are on the job full time and are paid full-time wages. In the reserves, you might serve part time or not at all. It depends which reserve force you are in.

All officers and enlistees sign up for a commitment of eight years. They may be on active duty for anywhere from two through eight years, depending on the branch they enter. However long they stay on active duty, the rest of the eight years are spent in the reserves. Some of the time may be spent in the Ready Reserve and some in the Standby Reserve, depending on the enlistment plan. Persons who spend eight years on active duty are exempt from reserve duty.

The Ready Reserve is composed of people who train one weekend a month and fifteen days sometime during the year. They are paid for each day spent in uniform. As the name implies, they are ready to be called up in the event of war or some sort of alert.

The Standby Reserve is made up of men and women who have left active duty yet still have some time left on their eight-year obligation. They do not have to attend training at all, and they do not get paid. The Standby Reserve can only be called up if Congress declares war or in the event of a national emergency.

The Retired Reserve is staffed by people who have put in enough years to qualify them for retirement benefits. They are paid nothing for being in the Retired Reserve.

History of the Reserves

This country has always had a reserve force in one form or another. The minutemen of the Revolution were an early version of the Ready Reserve. Various states have had militias that were trained and ready to respond to a call.

The Reserve Officers' Training Corps, or ROTC, was established in the years following World War I to secure a supply of trained officers. When the nation began mobilizing for World War II, the ROTC supplied close to one hundred fifty thousand officers to lead the rapidly swelling armed forces.

The second world war showed that the reserves were prepared to supply plenty of officers, but sadly lacking in enlisted personnel. Congress eventually passed a law requiring those who finished active duty to serve in the reserves. Since then it has become possible to join the reserves and not spend any time on active duty at all, except for the short period of basic and job training.

People who serve in the reserves come at a bargain price for the federal government. The reserves are much less expensive to maintain than the active forces, yet they are there if needed.

They were needed in Korea and Vietnam, and there was a time in the early sixties when international tensions caused President Kennedy to call the reserves into active duty.

Advantages of Joining the Reserves

Entering the reserves is a good way to tap into military benefits with a minimum investment of your time. If, for example, your goal in the service is to learn to repair computers, you can go through basic and computer school then return to civilian life and begin your career. If you are interested in a college education, you can get financial help through the reserves (see Chapter 15).

The biggest advantage of reserve duty is the money. In effect you have a second job, one that in the Army, for example, could pay as much as fifty thousand dollars over a twenty-year career—an average of twenty-five hundred dollars per year. In addition, you have other benefits such as low-cost life insurance, a limited right to buy things at discount military stores, recreational privileges on any nearby military installations, and possible payoff of any student loans you may have. You are also eligible for veterans' benefits such as low-cost mortgages, a pension, and payment of burial expenses. (Each branch of the service offers slightly different benefits, so it's best to check with a recruiter to get the latest information.)

Money aside, some people enjoy the chance to put on a uniform now and then and engage in training, perhaps in Europe or some place else appealing, the chance to exert leadership, and the chance to have good times with friends who come together monthly for reserve duty.

Disadvantages of Reserve Duty

If you are in the reserves, there is always the possibility that you will be called back into active duty. Given the way warfare has changed, the government is not likely to field a large fighting force as in World War II, but that is no guarantee your reserve duty will be limited to thirty-nine days a year. If, for example, the active duty units of your particular branch are called away to another part of the world, it's entirely likely that your unit could be activated to take their places until they return. And if the call comes, you have to go.

Being in the reserves can cause problems at work. Your employer may not be thrilled with your annual two-week training period, especially if you want to take a vacation in addition. If your firm has a big project to do during the time you are away, you may, depending on the work you do, lose valuable experience and face a poorer chance at promotion than a coworker with no military obligation.

As for the financial benefits of reserve duty, the more your salary rises in the civilian world, the less important your reserve paycheck will become. After a while the whole arrangement may seem more of a pain than it is worth. You may enjoy your access to military bases and the accompanying benefits if you are close to a base big enough to have a post exchange or movie theater, but lots of reserve units are nowhere near such installations.

Perhaps the biggest complaint of reservists is the monthly drills. These training days have an uncanny knack of coinciding with birthdays, anniversaries, and other important family occasions. If your local commander is fond of parades, for example, you may find yourself going left, right, left on a holiday such as the Fourth of July when you'd rather be somewhere else.

The National Guard

A special reserve force is the National Guard. This will be covered in Chapter 11, but for now the difference is simple: the reserves report to the federal government and are rarely called out for domestic problems. The National Guard reports to the federal *and* the state governments, and can be called out to assist in times of disaster or civil disorder.

OFFICER OR ENLISTED?

Another one of the big choices for someone joining the military is whether to go in as an officer or as an enlisted person. At first the choice might seem obvious—go in as an officer. You make more money and there are fewer people to tell you what to do.

It's actually more complex than that. Once more you have to consider what you expect out of the service. Do you plan to make it a career? Or are you using the military as a means of learning a job skill or getting money for college? Your answers to these questions will determine the path you should take.

Although each branch of the service does things differently, there are some things about officers and enlisted personnel that are the same in every branch.

Officers. An officer in the military is the equivalent of a corporate manager in civilian life. Over 90 percent of officers have completed four years of college, and officers enjoy benefits that enlisted personnel do not. Usually officers are the only ones who can take aviation training, and they are far more likely to be sent to graduate school or spend time in the civilian sector learning managerial skills.

The Army, Air Force, and Marines use the same titles for their officers: you start off as a second lieutenant and work your way toward becoming a general. Due to their nautical heritage, however, the Navy and the Coast Guard use different titles: you begin as an ensign and dream of the day when you become an admiral. To make comparisons easy, the Pentagon refers to rank by numbers. An ensign or a second lieutenant is an 0-1, and an admiral or a general is an 0-10.

Basic pay for officers is the same in every branch of the service. If you are an 0-1 just in the service, you are paid at the rate of $14,688 per year. After several months you are promoted to 0-2, and your pay goes up to $18,480. Once you reach 0-3 with four years of service, you bring home $25,668. If you stay in twenty years and reach the rank of 0-10—a four-star general or an admiral—you make $68,700 per year.

Basic pay is just that—the minimum you receive. Additional money may come for flight duty, sea duty, or hazardous duty. In addition you may receive various nontaxable allowances for food, housing, and travel or other expenses connected with your job in the military. Another source of money is reenlistment bonuses.

All members of the service, no matter what their rank or how long they have been in, receive thirty days of paid vacation per year. During basic training, they can get a leave only in the event of an emergency at home or elsewhere. The Red Cross has to verify that something dire has taken place.

After being commissioned, officers have to serve a certain number of years. Afterward they can resign—unless they have recently accepted a promotion, successfully completed flight training, or taken advantage of an educational offering. Since the military has invested more money in these cases, the officer is required to stay in longer.

Enlisted. An enlisted person in the military is equivalent to an hourly worker or a supervisor in the civilian world. The various branches have different names for the ranks, so to simplify things they are often referred to by numbers. An E-1 is the lowest rank; someone holding it would be called a private in the Army and the Marine Corps, an airman basic in the Air Force, and a seaman recruit in the Navy. Enlisted ranks go as high as E-9.

The first two or three promotions are more or less automatic, provided you progress satisfactorily. Although each branch of the service does things differently, you generally advance from E-1 to E-2 at the end of six months, and to E-3 at the end of a year. E-4 will take anywhere from one and a half to two and a half years. From there on up promotion is based on your performance on the job, on tests, on your commanding officer's recommendation, and on the needs of your particular branch.

Not everyone starts out at E-1. If you have a particular skill the service needs or have completed some college courses, you can sometimes enter at E-2 or E-3. In times of combat people often rise in rank much more quickly than normally.

Your pay is based on your rank, how long you have been in, and the situations in which you are serving. Special pay is given, for example, to those on sea duty or hazardous duty.

Let's focus on the Navy to get an idea of pay. When you become a seaman recruit (E-1), you are paid at the rate of $7,668 per year. After six months you are promoted to seaman apprentice (E-2) and draw pay at the rate of $8,592 per year. Whenever you make seaman (E-3) your pay goes up to $8,940—$9,420 if you've been in more than two years. If you make the Navy your career and work up through the ranks for twenty years to the position of Master Chief Petty Officer (E-10), you make $25,644 annually.

These pay rates look rather paltry when compared to civilian ones, but you've got to remember that for enlisted personnel the service provides food, uniforms, and housing. The food may be mess hall fare and the housing may not be the latest in design, but the price is right. If there is not enough housing or the base is too small to support a dining hall, you receive a monthly allowance so you can provide these things for yourself. If you are married and perhaps have children, the food and housing allowances are higher.

Most young people are single when they enter the service, and few have children when beginning a military career. The pay for entry-level enlisted personnel is sufficient for a single person, but trying to support a family on it is very difficult. Your spouse can work, of course, but the military will move you so often that he or she will have a hard time building a career. For a more detailed treatment of family life in the military, see Chapter 13.

Going in as an enlisted person does not mean you must always be one. There are several ways of making the jump from the enlisted ranks to the officer corps. You can apply to a service academy, enter an ROTC program, or go through officer candidate school. For further information on these programs, see Chapter 5.

As with officers, all enlisted personnel receive thirty days of vacation per year.

4. Going into the Enlisted Ranks

U.S. Air Force photo

ONE OF THE REASONS military people were so fond of the draft is that it simplified things. For the most part, young men were told when they had to leave, they reported to an induction center, and off they went. Now that no one is drafted, the military branches have come up with a sometimes bewildering number of ways to go in. They are constantly tinkering with these plans—adding bonuses here, expanding or shortening the length of hitch there—so the person who wants to enter the enlisted ranks has to look very carefully at the options. This chapter will cover the basic options in all branches and the three-step process of joining up: taking an aptitude test, talking with a recruiter, and going to a military Processing Station.

THE ARMED SERVICES VOCATIONAL APTITUDE BATTERY (ASVAB)

Most military applicants take the ASVAB before ever visiting a recruiter. This test, offered at many high schools, is much like college entrance exams such as the ACT or the SAT, but it focuses more on vocational than intellectual matters. The test gives the military a chance to see how many jobs you are likely to do well in. Your score will determine much of your future in the military, so it's important to do as well on this test as possible.

The test takes a total of two and a half hours to complete. It is in multiple-choice format: each question has four or five possible answers to choose from. The questions are grouped in ten categories:

- General science
- Word knowledge
- Numerical operations
- Auto and shop information
- Mechanical comprehension
- Arithmetical reasoning
- Paragraph comprehension
- Coding speed
- Mathematics knowledge
- Electronics information

Some people are dismayed at sitting down to what seems like a long and hard test. They sometimes find, however, that they do very well, for the test taps a lot of practical knowledge that is seldom asked for on high school exams.

If you are uncertain about the ASVAB, or fear that you won't do as well as you would like, it's a good idea to look at one of the ASVAB practice books now on the market. You might find one in a library or in a bookstore. These books list sample questions and give you practice at taking the test. According to recruiting officers familiar with the ASVAB, studying such books can raise your scores.

If you take the ASVAB and get your scores before going to the Processing Station, you can pick out a job with much more care and deliberation. You can ask a high school guidance counselor for a copy of the *Military Career Guide*. Mailed by the armed forces to all high schools, this book describes various jobs and estimates the ASVAB scores needed to qualify for them.

Comparing your scores with the job list is always interesting. You may find to your delight that you are qualified for the very thing you want to do. There's nothing like having a test confirm your hopes. Many people, however, are surprised at the jobs for which the test scores say they are qualified. A woman, for example, may find that she shows a strong aptitude for rigging parachutes, something she has never dreamed of doing. And a few people may find to their shock that they aren't qualified to do anything they like.

If you take the ASVAB early, before the day of processing you can ask questions of the recruiters, talk to people in the civilian world who have positions similar to those you're interested in, and—most important of all—think long and hard about your choice. The decision may still be hard, but at least you won't have to make it in fifteen minutes.

TALKING WITH A RECRUITER

A lot of people are confused about the role of a recruiter. Some think a recruiter will lie and do anything to get a young person to join the service. These people are half afraid that if they set foot in a recruiter's office they will get hit on the head and wake up at Marine boot camp with all their hair cut off. At the other extreme, some people assume that the recruiter knows everything about his or her branch of the military and will get the best possible deal for the person who comes in.

As usual in such matters, the truth lies somewhere in between. Recruiters do want to attract the right people to their branches, but they also work to make sure the recruit is satisfied. After all, no branch of the military wants a bunch of people who feel that they've been tricked.

Recruiters are friendly people. They drop what they are doing and offer you a chair. They smile and listen to what you have to say. Their courtesy and respect can come as a welcome change from the treatment you may have received from parents, teachers, or your boss. If you are having problems recruiters listen sympathetically. They offer you some coffee or a soft drink while you talk. They seem genuinely concerned.

At the same time you are looking at them. Whether male or female, recruiters generally look good. They wear clean uniforms with attractive ribbons and sparkling medals. Their shoes glisten. They seem happy with their work—or at least contented.

And when they start to talk, you listen.

You should also begin to think.

Let's say your uncle owns a used-car lot and offers to give you a car under some unusual conditions. You can have any car on the lot, but once you pick one you have to drive it and maintain it for three or four years. No matter how good or how bad the car turns out to be, it's all the transportation you will have for a long time. Under these circumstances you'll probably spend a long time making sure you get the best car. You should apply the same kind of thinking when talking to a recruiter.

The recruiter may show you a movie or a videotape depicting people in the particular branch having a wonderful time. The music will be stirring and the pictures inspiring. You may also see a film about basic training; this is so if you don't know what you are getting into you can head for the door. In addition, the recruiter will answer any questions you may have, and may tell you his or her own military story.

For the most part, recruiters like to talk about their careers. When you ask recruiters about their experiences in the military they will often tell you how good it is—and they will probably be telling the truth. Virtually every recruiter has been in the military long enough to reenlist. Someone who had a bad experience would not last that long.

Much like a car salesman, the recruiter is selling a product—military service—and like all good sellers he or she will accent the positive and play down the negative.

Basic Requirements for Enlistment

Besides giving you an overview of his or her particular branch, the recruiter will aim to quickly and efficiently discover if you will work out in the military. There was a time when the service would take almost anyone who walked in the door, but those days are long gone. Some people find to their astonishment that they cannot get in at all.

After some initial conversation, the two of you will get down to business. The recruiter will ask you questions about your background, testing your general knowledge and overall intellect and determining if you meet some basic requirements.

The recruiter will ask about your schooling, medical history, and interests. He or she will ask if you have a birth certificate and a Social Security card. If you were born overseas of American parents you need to show the recruiter papers that prove your citizenship. Aliens can enter the service, but they must provide proof of lawful entry for permanent residence.

The recruiter will also ask you about any drugs you have used or brushes with the law you have had. If you have a police record, you might as well admit it, for the recruiter will check with the police anyway. The recruiter will make sure you are at least seventeen years old, and if you are not yet eighteen you will have to get permission from your parents for the enlistment process to continue.

A high school diploma is not required except for women entering the Navy, Marines, and the Coast Guard. This doesn't necessarily mean that if you are a high school dropout you will automatically be accepted. If a particular branch of the service has many people trying to get in, it may choose those with diplomas over those without them.

If you have not yet taken the ASVAB, the recruiter may ask you to take the Enlistment Screening Test, a forty-five-minute test consisting of forty-eight questions covering math, vocabulary, and paragraph meaning. This test does two things. First, it gives you a sense of what the longer ASVAB is like. Second, it gives the recruiter a sense of what *you* are like. If you utterly fail the test, he or she will gently suggest that you and the armed forces would not work out well together.

If you have not taken the ASVAB, again, you may next be asked to take a practice version. This is another means of narrowing the field of applicants. It is given in the recruiter's office, at your first or a subsequent visit, and it also takes about forty-five minutes. If you score high enough you get a green light to proceed with the enlistment process, and you will probably be asked to begin filling out forms. The recruiter will also start to talk to you about jobs that you may qualify for.

Be a Good Consumer

Here are some things to keep in mind when you visit a recruiter.

Don't rush into anything. Your fiancé(e) may have told you to drop dead and you may be sick of your job, but don't rush off and recklessly join the military just to get away. Eight years can be a long time to regret your impulsiveness.

Make several trips. You are planning to enter an agreement that will cover almost one-fifth of your working life; surely you have enough time to do it right. The various enlistment options are sometimes hard to understand at once. An Army sergeant who says "You take the ASVAB to determine your MOS before going to basic and AIT" means that you will take an aptitude test to determine what job you qualify for. Then you go to basic training and job training. Spend a lot of time with the various recruiters so that you understand exactly what they are talking about.

Recruiters generally won't say anything bad about their own outfits, so ask them about the others. A Navy recruiter can be very frank about the shortcomings of the Army and the Air Force. The Marines will gladly point out why you shouldn't join the Army or the Navy, and the National Guard representative will fill you in on all of the others.

Talk to other people. Don't get all your information from a recruiter. Talk to someone who has been in the service. Best of all, talk to someone who is still in uniform. If you don't know such a person ask your friends and relatives for names of people to contact. Christmas and other holidays are good times to catch military people at home.

If you are black or Hispanic, make a special effort to talk to blacks or Hispanics who have been or still are in the service. A white recruiter has little grasp of the problems that face members of minority groups in uniform.

The same goes for women. Recruiters and their brochures don't always emphasize the limitations that women encounter in the service.

Note: Keep in mind that ex-servicemen and -women don't always tell the whole story either. The young man from your neighborhood who was thrown out of the Navy for drinking or drug use on a ship may not give you an accurate view of his military experience.

Take along a parent or a friend. Once you have gathered the basic information, it's a good idea to take someone along when you begin to ask questions. The recruiter is probably older than you and may sometimes be intimidating sitting there in uniform. Having an older person along may help in asking the tough questions, and your companion may think of things that you forgot to ask.

Take a list of the questions you want to ask. While you and the recruiter are talking your companion can take notes.

If you happen to run into an unethical recruiter, he or she will think twice about lying to you when someone else is writing things down.

When you feel satisfied with the answers to your questions—and the recruiter feels satisfied with *you*—it's time to schedule your appointment at a Military Entrance Processing Station.

THE MILITARY ENTRANCE PROCESSING STATION

All of the armed forces except the Coast Guard use the same facilities to prepare young people for entry into the military. A station is usually located in a city or large town, and you may get there by bus, plane, train, or car—all at the expense of the branch you are thinking of joining. If the facility is far from your home, you may be put up in a hotel overnight. At the Processing Station the personnel will check certain papers you'll have been asked to bring along, give you the ASVAB if you haven't already had it, take your medical history, and give you a physical examination. Next you may sign up for a special enlistment program, and you'll choose a job and entry date. Then it's time for swearing in.

Note: The personnel at the Processing Stations are all too familiar with applicants who have a wild party the night before they come in. These people are often sick and otherwise hung over. They do poorly on written tests, cannot remember medical histories, and are sometimes too incapacitated to take the physical. The party may be memorable, but when you wind up taking the bus instead of the oath, it somehow doesn't seem worthwhile.

Physical Requirements for Enlistment

The possibility of being in combat colors most of the policies and practices of the military. Equipment and uniforms are selected for the worst possible conditions they are likely to experience. The same can be said for people in the armed forces. The authorities try to make sure that everyone who gets in is healthy enough to perform under the worst conditions. There is no time during an amphibious assault, for example, to deal with the special problems of a diabetic. And a submarine, which can cruise under water for six months, is not likely to return to port because someone's allergies are acting up. For these reasons an ailment or deficiency that poses little or no problem in civilian life can be enough to keep you out of the armed forces.

The requirements for getting into the service are much the same across the board, although each branch has special requirements for special types of recruits. Standards for officers are different from those for enlisted personnel. Special programs such as aviation training or submarine duty have tougher standards. You won't, for example, see a 6-foot-6-inch fighter pilot; he wouldn't fit in the aircraft.

The military has a long list of conditions that will prevent you from joining up. It has another list of conditions for which you can apply for a waiver. By granting a waiver they acknowledge that you have some sort of problem but decide to take you anyway. One waiverable condition is height. In all the services the minimum height is five feet. You might be a particularly attractive person—with high ASVAB scores or a skill the military needs—yet you're only four feet, ten inches tall. So you get a waiver and they take you anyway, with no plans to put you on the barracks basketball team.

These waivers are often dependent on supply and demand. When enlistments are running high, perhaps no waivers will be given. However, when a particular branch is not getting many people, or not many skilled people, it will often take applicants with waiverable conditions. Oftentimes this happens toward the end of the month when the military is feeling the pressure to meet the quota. If you have a condition that you think might cause you problems, then your best bet would be to go in at the end of the month.

Conditions that will keep you out of the military. The following are some of the conditions that will prevent your entering the military. You cannot get a waiver for these problems. For a more complete list, and one that reflects the special requirements of a particular branch of the service, you should check with the appropriate recruiter.

The conditions include:

Drug addiction	Chronic alcoholism
Use of a hearing aid	Braces on your teeth
Diabetes	Drug therapy
Severe harelip	Imbecility
Multiple sclerosis	Muscular dystrophy
Obesity	Malignant tumor
Epilepsy	Severe bee sting allergy
Wool allergy	Severe allergies in general

Conditions for which you can apply for a waiver. The following conditions can also keep you out of the military, but for these you can apply for a waiver. They might be illnesses that you suffered five or ten years ago but have had no recent problems with. If you have had any of these conditions you should supply documents from your physician that explain the extent of the condition or the lack of recent problems. This is not a complete list; for complete details you should check with the recruiter for the specific branch you are interested in.

Hepatitis	Deformity or loss of fingers or toes
Stomach ulcer	Somnambulism (sleepwalking)
Foot trouble	Ear surgery
Hearing problem	Surgery on female organs
Bone or joint surgery	Absence or interruption of menses
Asthma	Healed fractures with any plates, pins, rods, or other device in place
Psoriasis	
Enuresis (bed-wetting)	
Hernia surgery	
Back trouble	

Height and weight requirements. The maximum height for men and women is 6 feet, 6 inches; the minimum is 4 feet, 11 inches for men and 4 feet, 10 inches for women. Heights of 4 feet, 10 inches or 4 feet, 11 inches are waiverable.

Weight Standards for Men

Height (in.)	*Minimum (regardless of age)*	*16–20 Years*	*21–30 Years*	*31–35 Years*	*36–40 Years*	*41 Years and Over*
58	98	147	153	151	147	140
59	99	152	157	156	152	145
60	100	158	163	162	157	150
61	102	163	168	167	162	155
62	103	168	174	173	168	160
63	104	174	180	178	173	165
64	105	179	185	184	179	171
65	106	185	191	190	184	176
66	107	191	197	196	190	182
67	111	197	203	202	196	187
68	115	203	209	208	202	193
69	119	209	215	214	208	198
70	123	215	222	220	214	204
71	127	221	228	227	220	210
72	131	227	234	233	226	216
73	135	233	241	240	233	222
74	139	240	248	246	239	228
75	143	246	254	253	246	234
76	147	253	261	260	252	241
77	151	260	268	266	259	247
78	153	267	275	273	266	254
79	157	273	282	280	273	260

Weight Standards for Women

Height (in.)	*Minimum (regardless of age)*	*17–20 Years*	*21–24 Years*	*25–30 Years*	*31–35 Years*	*36–40 Years*	*41 Years and Over*
58	90	121	123	124	126	135	135
59	92	123	125	129	129	139	138
60	94	125	127	132	132	142	141
61	96	127	129	135	136	145	147
62	98	130	132	139	141	148	147
63	100	134	137	141	145	151	150
64	102	138	141	145	150	156	154
65	104	142	145	149	155	161	159
66	106	147	150	154	160	165	164
67	109	151	155	159	165	171	169
68	112	156	159	163	169	176	174
69	115	160	164	168	175	181	179
70	118	165	169	173	180	186	184
71	122	170	174	178	185	192	190
72	125	175	178	183	190	197	195
73	128	180	183	188	195	202	200
74	132	184	189	193	201	208	206
75	136	189	194	199	206	214	212
76	139	195	199	204	212	219	217
77	143	200	204	209	217	225	223
78	147	205	209	215	223	231	229
79	151	209	213	219	227	234	231

Waiverable Weights for Men and Women

	Men		*Women*	
Height (in.)	*Minimum*	*Waiverable to*	*Minimum*	*Waiverable to*
60	100	90	94	85
61	102	92	96	86
62	103	93	98	88
63	104	94	100	90
64	105	95	102	92
65	106	95	104	94
66	107	96	106	95
67	111	100	109	98
68	115	104	112	101
69	119	107	115	104
70	123	111	118	106
71	127	114	122	110
72	131	118	125	113
73	135	122	128	115
74	139	125	132	119
75	143	129	136	122
76	147	132	139	125
77	151	136	143	129
78	153	138	147	132

Enlistment Programs

The branches of the armed forces change their enlistment programs from time to time. When the number of volunteers is down throughout a branch or in certain jobs, the branch may offer bonuses or special programs designed to lure the needed personnel. Similarly, when lots of people are signing up bonuses may be lowered or cut entirely, and certain programs may be discontinued.

For this reason it is impossible to list all the current options here. *A Young Person's Guide to Military Service* can give you an idea of what is offered, but to find out about current programs you should check with the recruiters for each branch.

Following are enlistment programs that are offered by all branches. The names are not always the same, and certain details may differ, but in general the programs are uniform.

Enlistment Programs in the Active Forces

□ *Delayed Entry Plan.* In all of the branches you can join now but put off entering until there's an opening for the training and the job you want. You can wait as much as one year between taking the oath and reporting for duty.

□ *Stripes for Skills* or *Stripes for College.* This option benefits people who have a skill or college experience that the service needs. Typically, you go in at the rank of E-2 or E-3 instead of the usual E-1. You thus make more money and are further ahead for promotions.

□ *Buddy Plan.* If you go in with a friend you can undergo basic training together, begin duty at the same station, or both. Given a choice you should pick the latter, for during basic training you and your friend could be in different companies and see very little of each other.

□ *Base of Choice* or *Country of Choice.* If you choose your job from a certain list you can pick your base or country. (Ask how long you can stay in those places.)

Enlistment Programs in the Reserves

□ *Split-training Option.* This is for high school or college students who do not want their boot camp and job training to interfere with their schooling. You can go to basic training one summer and job training the next.

□ *Stripes for Skills* or *Stripes for College.* Same as in active duty.

Selecting a Job: Questions to Ask

Assuming you pass the physical, the most important thing you'll do at the Processing Station is sit down with a job counselor to decide what field you will enter. That list of hundreds of jobs you saw in the recruiter's office will, depending on your exam scores, be narrowed considerably. And you won't have much time to choose among the jobs open to you; a counselor usually spends fifteen to thirty minutes per recruit.

This is a crucial point for you. The counselor, or classifier, as the military calls this person, may pressure you to make up your mind quickly. He or she may want to go to lunch or move on to the next person. Don't let someone intimidate you into making a hasty decision. You haven't signed anything and no one can order you around—not yet, anyway. The people at the Processing Station may act irritated with you for taking a lot of time, but that's their problem, not yours. You're probably never going to see them again anyway. Here are some good questions to ask:

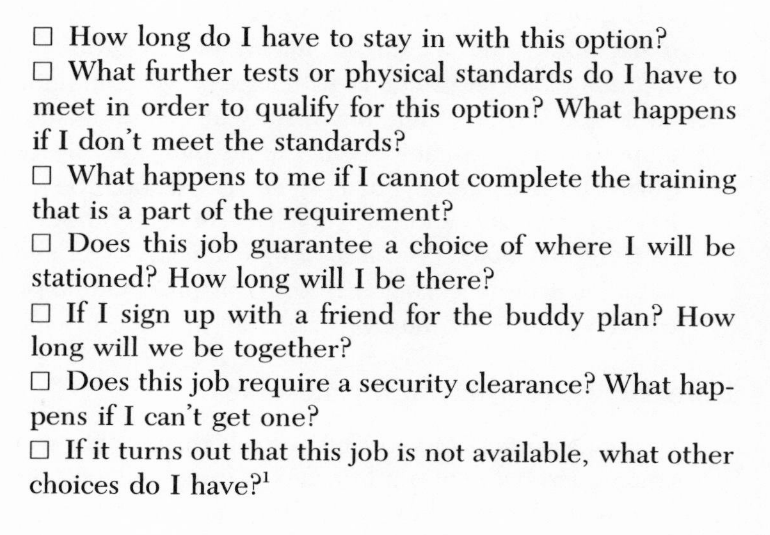

☐ How long do I have to stay in with this option?

☐ What further tests or physical standards do I have to meet in order to qualify for this option? What happens if I don't meet the standards?

☐ What happens to me if I cannot complete the training that is a part of the requirement?

☐ Does this job guarantee a choice of where I will be stationed? How long will I be there?

☐ If I sign up with a friend for the buddy plan? How long will we be together?

☐ Does this job require a security clearance? What happens if I can't get one?

☐ If it turns out that this job is not available, what other choices do I have?[1]

Once you choose a military specialty, the classifier will check a computer to see when the next opening for that job will occur. Your training will take place in time for you to fill the open slot. Depending on the job, you might have to wait up to a year to go in. Some people who are eager to join will choose the specialty that will get them into uniform the quickest. This is not always a wise choice. You may be doing this job for four to six years; make sure it is one you will be happy with.

---

1. *Nine Things to Remember* (Cambridge, Mass.: American Friends Service Committee).

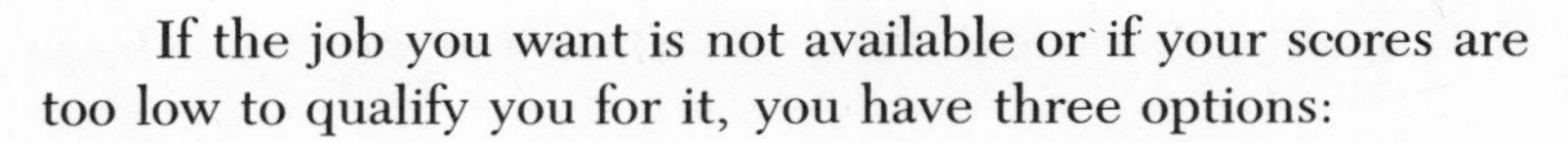

If the job you want is not available or if your scores are too low to qualify you for it, you have three options:

- ☐ *One,* you can ask for a waiver. If the required score is an 86 and you have a 76, ask for special permission to take the job anyway. The classifier may not be able to grant this, but the commander of the Processing Station can. Don't hesitate to ask; the commander may be impressed with your determination and give you what you want.
- ☐ *Two,* go home and wait for several weeks to see if the situation changes. Jobs come and go on the availability list, and the qualifying scores for various jobs float up and down depending on how urgently the military needs people. What you are offered one week may change completely the next. You can always come back to the Processing Station; your physical is good for thirty days.
- ☐ *Three,* go home and call your recruiter. Explain the situation, and let that person fight for you. The recruiter's job is to get you into uniform, and he or she can monitor the job situation and call you when things improve.

Don't be afraid to go home without having settled on a job. It's nice to wrap up everything in one trip, but your overriding concern should be getting a job you will be happy with. Once you are home and away from all those people in uniform, you'll be better able to think through your options.

Final Steps

Once you do choose the job and the accompanying date of entry to the service, you will complete your application for enlistment, including something called a guarantee statement. If the classifier makes any specific promises to you, make sure they are written on this statement. Spoken promises are worthless.

Finally, all your papers will be checked. Now you are ready to take the oath of enlistment. Before you do, someone will read you the oath, explain what it means, and describe the penalties that will occur if you take it and then try to back out.

It's not always easy to think at the Processing Station, but now is the final time to consider whether or not you wish to enter the military. Your signature and the oath of enlistment will profoundly affect the next eight years of your life. Make sure you know what you are getting into.

Assuming you decide to join, you then go to the Ceremony Room, where you and any other recruits present will raise your right hands and officially enter the service. The oath of enlistment is the same for all the branches:

> I do solemnly swear that I will support and defend the Constitution of the United States against all enemies, foreign and domestic; that I will bear true faith and allegiance to the same; and that I will obey the orders of the President of the United States and the orders of the officers appointed over me, according to regulations and the Uniform Code of Military Justice. So help me God.

You will have just joined the reserve portion of your branch of the service. At this point a few people head right off to boot camp, but most go home to wait until a slot opens for them. For some this time cannot pass quickly enough; they are eager to get going. Others savor their remaining days of leisure by sleeping late. A few people realize that they have made a mistake and want out. Chapter 12, How to Get Out, deals with the last case.

When the time comes you return to the Processing Station, where you undergo a quick physical exam, an interview, and another swearing in. This second oath transfers you from reserve to active duty. Then it's off to boot camp, where you begin your enlisted military career.

5. Going into the Officer Corps

Official U.S. Navy photograph

Whether you plan to make the military your career or not, there are good reasons for going in as an officer. You will receive more pay, more responsibility, and a greater opportunity for advancement. Officers usually gain managerial experience and advance more rapidly than do their civilian counterparts. And the officer corps offers you the most opportunities to further your education at low cost. This, plus the experience you gain while in uniform, can be very useful when you leave the military and apply for a civilian job.

After discharge from the service, officers often earn exceedingly good civilian salaries. Many firms that work with the Pentagon are eager to hire former officers who know the military, often have a security clearance, and may have valuable military connections.

The physical standards for officers are the same as those for enlisted personnel (see Chapter 4). The educational standards, however, are different. All incoming officers must have had at least two years of college.

Most officers start off at the lowest rank—second lieutenant in the Army, Air Force, and Marine Corps, or ensign in the Navy and Coast Guard. All the branches but the Coast Guard offer three ways of becoming an officer: through a service academy, through the Reserve Officers Training Corps (ROTC), or through an officer candidate school (OCS). A fourth means is by direct appointment, but this is usually reserved for men and women in professions such as law, medicine, and the clergy.

THE SERVICE ACADEMIES

If you are thinking about becoming an officer, then the four service academies are the best place to begin. Whether they come from the Military Academy (West Point), the Naval Academy (Annapolis), the Air Force Academy, or the Coast Guard Academy, graduates of these elite schools are considered the cream of the officer corps.

American history is illuminated with the names of men who attended service academies. Robert E. Lee, Ulysses S.

Grant, Douglas MacArthur, Dwight Eisenhower, George Patton, Chester Nimitz, Hyman Rickover, and Jimmy Carter are some of the generals, admirals, and presidents who once wore cadet uniforms.

If you are accepted to one of the academies you get four years of college for free, plus pay, medical care, and room and board. Unlike your counterparts in ROTC, you undergo an entire curriculum that is aimed toward military service. Once you graduate you are given first crack at choice assignments such as aviation and submarine duty.

Even more important, the academies form the ultimate "old boy network" in the military—a network that will swing into action when promotions and assignments are up for grabs. It's no wonder that academy graduates, having had helping hands along the way, are the top leaders in almost every branch of the service.

The academies emphasize the strong character and leadership qualities of their graduates and constantly stress that cadets are very special people. In one sense they *have* to be special people; they put up with a system in effect at only a handful of colleges in this country.

First of all, it is very difficult to get in. A complicated process pits applicants against each other in securing nominations from a variety of high governmental officials. Once a young person receives a nomination, he or she may have to survive another round of competition to get in. Academic records, athletics, and the all-important leadership qualities are scrutinized in deciding who gets in.

The only exception to this is the Coast Guard Academy, which has no appointments, quotas, or special admissions categories. Everyone who wants to get in participates in one competition.

Once in—no matter which academy—it's no picnic. Freshmen are subjected to a degree of harassment that would never be tolerated at a conventional school. The physical and psychological pressure is intense, much as in boot camp, except it lasts for the entire first year. At West Point alone, over 35 percent of the students fail to graduate.

Cadets must adhere to an honor code that requires them to leave for the slightest infraction of the rules. Cadets have been removed from academies for suspecting others of breaking the rules and not reporting it to their superiors. In addition, all cadets must participate in athletics—like it or not—on an intercollegiate or intramural level.

The academic training at a military academy concentrates on "the art and science of warfare" and thus is more limited than that of a conventional college. The faculty, though they be high-ranking officers, rarely have the academic credentials that one might find among professors at an equally selective civilian college.

After graduation the newly commissioned officers are faced with an obligation for five years of active duty—the longest of any beginning officers.

After discharge from the service, officers often earn exceedingly good civilian salaries. Many firms that work with the Pentagon are eager to hire former officers who know the military, often have a security clearance, and may have valuable military connections.

If you qualify to get into an academy, then you could stand an equally good chance of getting into an Ivy League school or a prestigious university such as MIT or Stanford. All of these schools have extensive financial aid programs, so cost should not be the sole determining factor in your decision.

Perhaps the best way of determining whether you would like to go to an academy is to visit one or more of them. Actually walking the grounds and talking with cadets can reveal the good points and the bad in a way that no brochure or book ever can. All of the academies encourage visitors, especially those who are considering applying for a nomination.

In addition, all of the academies have liaison officers around the country who can answer your questions and guide you through the complicated application procedure. You can get the name of the officer nearest you by writing the appropriate academy.

Brochures for the four academies clearly point out that cadet life is not for everyone. Despite this, thousands of high school students apply every year, and the academies are never at a loss for qualified students from whom to choose. If you find out too late that a service academy is no place for you, you can resign during the first two years and incur no military obligation.

West Point. The United States Military Academy is located on sixteen thousand acres of land on the West Point of the Hudson River, about fifty miles north of New York City. It was established in 1802, making it the oldest of the service academies. The first class had ten cadets.

Now West Point has approximately forty-five hundred cadets. Each receives a general education in mathematics, science, engineering, English, history, social sciences, national security, and psychology. Cadets have a choice of two tracks of study: a mathematics-science-engineering track and a humanities-public affairs track.

During the summers cadets master such things as artillery firing, mountaineering, wilderness survival, and tank operations. They may take jungle warfare training in Panama or northern warfare training in Alaska; they may learn to pilot a helicopter or use a parachute.

Annapolis. The United States Naval Academy is located on 329 acres of land in Annapolis, Maryland, on the Chesapeake Bay, about ten miles from Washington, D.C. It was established in 1850.

Annapolis has forty-five hundred midshipmen, as cadets are known. It trains officers for both the Navy and the Marine Corps. "Middies" study mathematics, science, social studies, and the humanities. At least 80 percent of the midshipmen must major in engineering, math, or science-oriented disciplines. The rest can major in the humanities or social sciences.

In the summer after their first year of study midshipmen go to sea with fleets in the Atlantic, Pacific, or Mediterranean. There they become familiar with life at sea and assume some leadership positions. The following summer they may participate in aviation, submarine, or Marine Corps training, and the next year they go back to sea and practice celestial navigation, weapons training, and basic fleet tactics.

Air Force Academy. The newest of the military academies is located on eighteen thousand acres of land near Colorado Springs, Colorado. It opened in 1955 with a class of 306 cadets.

Now there are 4,406 cadets at the Air Force Academy—3,884 men and 522 women. They take courses in science and engineering as well as in the social sciences and humanities. There are specific majors in each field, yet cadets can elect a nonmajor path through a wide variety of courses.

Summers find cadets taking training in survival, evasion, resistance, and escape in the Rocky Mountains. Later they can serve at Air Force installations to experience and observe life in the active Air Force and participate in activities such as soaring, parachuting, light-plane flying, and navigation training.

Coast Guard Academy. The smallest of the service academies, the Coast Guard Academy is located on one hundred acres at the mouth of the Thames River in New London, Connecticut. It was founded in 1877, and it now accommodates nine hundred cadets.

The cadets have a choice of nine majors: civil, electrical, marine, or ocean engineering; marine, mathematical, or physical science; government; or management.

During the summers cadets take a long cruise on the *Eagle,* a three-masted sailing ship, or on a Coast Guard cutter. They are trained in seamanship, navigation, damage control, and fire fighting. They assist in day-to-day operations at Coast Guard installations and sometimes take cruises to foreign ports.

Eligibility

To qualify for admission to any of the service academies you must be a U.S. citizen who is at least seventeen years old but not yet twenty-two years old on July 1 of the year you plan to enter. You cannot be married or plan to marry during your four years at an academy, and you cannot have any dependents. In considering whom to accept, the admissions staff looks at the whole candidate, taking into account academics, physical fitness, and leadership qualities.

Since all of the academies lean toward science and engineering majors, applicants are urged to take four years of math and plenty of science in high school. Students may take either the American College Testing Program (ACT) test or the Scholastic Aptitude Test (SAT) of the College Board Admission Testing Program. And it always helps to have good grades.

Sports and physical conditioning are a big part of academy life; students must participate in intercollegiate or intramural sports. Anyone seeking a nomination must undergo an extensive physical examination—one with the highest standards in the military. Participation in varsity or other sports in high school is very helpful in getting in. Out of the 1,330-member class of 1990 at West Point, for example, 1,158 cadets won an athletic letter while in high school.

Obviously, people who plan to become officers should be good leaders. Each academy looks closely at how many offices you held in high school organizations or how many teams you captained. Participation in scouting, church, or civic activities is also considered.

Because of the academies' whole-candidate approach, it is possible that an applicant with low grades but with very high leadership potential can still get in. The same goes for someone who has perfect grades but is weak in athletics. As with all colleges, fielding winning intercollegiate teams is important. If you are an All-American pass receiver or a winning quarterback, you may find it downright easy to get into a service academy.

Obtaining a Nomination

Unless your father was a Medal of Honor winner in one of the services, you will have to secure a nomination from an elected or military official to be considered for admission to any service academy except the Coast Guard's.

Admissions officers at the academies urge that interested high school students apply for nominations no later than the spring of their junior year. You should apply in all of the categories for which you are eligible. Congressional appointments are the most important; by law the first 150 appointments to any academy and three-fourths of the remainder must be congressional nominees.

You do not have to know your senator or representative personally to obtain a nomination. Most elected officials diligently seek the best qualified candidates from their district. Politics, however, sometimes enters these nominating procedures. If your mother was a campaign treasurer for someone who tried to defeat your representative, then you might have a hard time securing a nomination from that particular official.

If you have your sights on a service academy, it's not a bad idea to approach your senator or representative before it is time to ask for the nomination. You might do this by writing the official a letter, working in a campaign for the official or the party, or visiting the official's local or Washington office. It can't hurt.

Once you've determined you are eligible and decide you want to go, you should apply to one or more of the following:

☐ *U.S. senators and representatives, the delegate from the District of Columbia, and the resident commissioner of Puerto Rico.* These officials can have five men or women attending each of the academies at one time. If all five spots are filled, no nominations can be made until someone graduates and a place opens up. For each vacancy, ten persons can be nominated. You should apply to the appropriate official for a nomination.

☐ *The president of the United States.* The president can make up to 100 appointments a year to each academy, but the applicants must be the children of career members of the armed forces (including the Coast Guard) who are on active duty, retired, or deceased. You should apply to the superintendent of the appropriate academy, and not to the president directly.

☐ *The vice president of the United States.* The vice president can have five people from the country at large attending each of the academies. Whenever there is a vacancy from one of these spots, the vice president can nominate ten applicants. This is the most competitive nomination, and you should apply to the Office of the Vice President, Washington, D.C., 20510 by September 1 of your senior year in high school.

☐ *The governor and resident commissioner of Puerto Rico; the governors of the Canal Zone and American Samoa; and the delegates from Guam and the Virgin Islands.* Each of these officials may have one person in attendance at each academy. The governor of Puerto Rico nominates only natives of the island, whereas the resident commissioner of Puerto Rico nominates any residents of the island. In each case you should apply to the officeholder directly.

If you fall into one of the following categories, you need not secure a nomination from an elected official.

☐ *Regular Army, Navy, Air Force, Marine Corps personnel.* They are eligible if they have completed one year of active duty by July 1 of the year they wish to enter an academy. They should apply to the commanding officer to compete for 85 appointments at each academy.

☐ *Reserve Army, Navy, Marine Corps, and Air Force personnel.* These applicants must be on active duty or belong to a drilling unit, and they must have served in the reserve for at least one year prior to July 1 of the year they wish to enter. They apply to commanding officers to compete for 85 appointments to each academy.

☐ *ROTC members.* These applicants compete for ten appointments at each academy. They should apply through their professor of military science.

☐ *Students at honor naval and military schools.* These military prep schools will be discussed later in the chapter. Each headmaster nominates three applicants, who compete for ten appointments at each academy. Apply to the appropriate headmaster.

- ☐ *Children of deceased or disabled veterans and children of prisoners of war or persons missing in action.* Children of military personnel who were killed in action or who died or were totally disabled from wounds, injuries, or disease received while on active duty are eligible, as are the children of servicemen or civilians who are prisoners of war or missing in action. Apply to the superintendent of the intended academy. A maximum of sixty-five appointees may be at one academy. Anyone who is eligible for this category is barred from competing for the presidential nominations.
- ☐ *Children of Medal of Honor winners.* These applicants have the best chance of all, for if they are qualified they are automatically admitted, and there is no limit to the number of positions available to them.

The vice president, congressmembers, governors, the delegate from the District of Columbia, and the commissioner of Puerto Rico can use any method to select their nominees. Some use civil service screening examinations. Most use some variation of the following three methods.

- ☐ *Competitive method.* This is the most commonly used method and the one the military prefers. The official submits a list of nominees to each academy. The academies examine the records of each nominee and rank them in order of preference. The person ranked highest gets the principal nomination.
- ☐ *Principal-alternate (noncompetitive) method.* In this process the official makes the selection, choosing one person as the principal nominee and ranking the rest as alternates.
- ☐ *Principal-alternate (competitive) method.* Here the official picks his or her number-one nominee and lets the academies rank the rest.

Just because you are not the principal nominee does not mean you should give up. After all of the reserved spots are filled, the academies will dip into the ranks of the alternates to make up the remainder of the entering class. So even if your congressmember has reached his or her quota for a particular academy, you still have a chance.

And there's always next year. If you can wangle another nomination the academies will consider you once more. For this you need to update your files at the academies to which you have applied. In the meantime, you may decide to go to a prep school.

Military Prep Schools

The Army, Navy, and Air Force operate prep schools for people who would like to go to the academies but feel they need one more year of schooling to meet the admission standards.

Students at these schools fall into two main groups. The first group are men and women who are already in either active or reserve units of the armed forces. They may have been out of high school for a while and need to do something to get back in the academic swing.

The second group is composed of some of the more qualified civilians who received a nomination but were not chosen by the academies. This additional year of schooling gives them an opportunity to improve their academic records, participate in athletics, and demonstrate their leadership abilities.

Getting into one of the prep schools does not mean acceptance into one of the academies will follow automatically. The next year you have to secure another nomination and go through the competition again.

Various branches of the service offer preparatory scholarships to selected prep schools and junior colleges throughout the country. This extra year of schooling can serve the same purpose as a year spent at one of the military prep schools, and it may even be more helpful.

U.S. Military Academy Preparatory School. The Army operates its prep school at Fort Monmouth, New Jersey. Military personnel and civilians who wish to attend must request admission to the school. Write the Commandant, U.S. Military

Academy Preparatory School, Fort Monmouth, New Jersey 07703.

U.S. Naval Academy Preparatory School. The Navy and Marine Corps maintain a prep school in Newport, Rhode Island. Military personnel and civilians do not apply for this school; the Navy looks over the nominees who have been turned down by the academy and offers selected ones positions at the prep school. Further details about the school can be obtained by writing the Director of Candidate Guidance, Box C, U.S. Naval Academy, Maryland 21402.

The U.S. Naval Academy Foundation offers scholarships to nominees who wish to spend a year at a participating prep school, junior college, or college selected by the applicant. Scholarship applications may be obtained from the Executive Director, Naval Academy Foundation, 48 Maryland Avenue, Annapolis, Maryland 21401. Applications should be received by April 1 of each year.

A similar opportunity for enlisted men and women is offered in the Broadened Opportunity for Officer Selection and Training program, or BOOST. Participants in BOOST receive up to twelve months of academic preparation so they will be better able to compete for an academy appointment or a Naval ROTC scholarship. The school is in San Diego, California.

BOOST is both an opportunity and a gamble. For those already in the service, it is an opportunity to rise from the enlisted to the officer ranks. For civilians it's more of a gamble. To get into BOOST you have to join the Navy first. The Navy may guarantee in your enlistment papers that you'll get the preparatory schooling, but they won't guarantee that you will either get into the academy or receive an ROTC scholarship. And if you don't, you spend the rest of your eight-year obligation in the enlisted ranks of the Navy.

The Air Force Academy Preparatory School. The Air Force prep school is located approximately five miles south of the Air Force academy in Colorado. Civilians who have been nominated but have not received an appointment to the academy are automatically considered for admission to the school, but military personnel are considered only if they apply to the school at the same time they request a nomination to the academy. They usually do so, as a way of hedging their bets.

The Air Force has three scholarship funds that help academy nominees to attend other schools. The Falcon Foundation makes annual cash grants for students to attend specific prep schools in various parts of the country. Applications and information can be obtained by writing the Falcon Foundation, P.O. Box 67606, Los Angeles, California 90067. Applications must be received by May 1 of the year you wish to attend.

The Gertrude Skelly Trust Fund offers scholarships only to the children of active, retired, or deceased career members of the armed forces. Applications and information can be obtained by writing the Gertrude Skelly Trust Fund, P.O. Box 1349, Tulsa, Oklahoma 74101. Applications must be in by May 1 of the year you wish to attend.

The Air Force Aid Society sponsors the General Henry H. Arnold Educational Fund, which provides scholarships for the children of Air Force personnel. Applicants can make their own choice of accredited schools. Write for details to the Director, Air Force Aid Society, National Headquarters, Washington, D.C. 20333. Your application must be in by January 31 of your senior year in high school.

The Coast Guard does not have a prep school of its own, but it does reserve fifteen places in the U.S. Naval Academy Preparatory School. These places are offered to outstanding candidates who did not get into the Coast Guard Academy. The Coast Guard also offers ten slots in the Navy's BOOST program.

The Coast Guard Academy Admission Procedures. The Coast Guard has the easiest application procedure; nominations from elected officials are unnecessary. Instead, applicants to the Coast Guard Academy participate in an annual nationwide competition that is based on high school ranking, performance on the ACT or the SAT exam, and leadership potential.

According to the Coast Guard *Bulletin of Information*,

each applicant is judged on his or her "academic background, the possession of aptitudes related to both technical and cultural studies, a sincere interest in the Coast Guard as a career, and relevant personality and physical characteristics." The academic portion of the requirements carries a 60 percent weight in the determination, and everything else carries a 40 percent weight.

When all this was put into practice for the class of 1990, approximately 4.6 percent of the applicants got in.

RESERVE OFFICERS' TRAINING CORPS

Most of the officers in the armed forces come through ROTC These men and women attend public or private colleges and take military training alongside their other courses. After graduation they are commissioned into the officer ranks and are obligated to serve for either three or four years, depending on the amount of scholarship money they received from the government.

ROTC versus the Academies

Whereas graduates of the service academies may have the inside track to flight training and promotions, a strong case can be made that ROTC is the better path to follow. Although neither as competitive as the academies nor offering all their benefits, ROTC gives its members a chance to be part of both the military and the civilian worlds during their college years. By wearing a uniform and participating in classes and drills, the student gets a chance to sample military life before making the big commitment. And doing this in a normal college setting has distinct benefits from the point of view of both the military and the individual.

From the military perspective, ROTC members who are surrounded by civilians have to frequently justify their decision to join the service, and this forces the prospective officers to think out and articulate their beliefs under sometimes hostile questioning. Late-night dormitory exchanges make them clarify their reasons for joining the service and defend these reasons in a way that cadets at a military academy seldom do. Like a piece of steel that has been repeatedly heated and plunged into water, an ROTC candidate goes through testing that produces an officer whose commitment to military service is unshakeable.

From the individual's point of view, the civilian influence on ROTC candidates is important. Through contact with professors and fellow students, future officers hear ideas and points of view that are not often spoken in the more rigid academy setting. They may come away from this larger marketplace of ideas with a broader view of the world and its various peoples.

Another important consideration for the individual concerns bailing out of the military. Your college years are a time of finding things out about yourself, sampling various disciplines, and hunting for your niche in life. If you decide to drop out of ROTC—before you have incurred a military obligation—it's not such a big deal. You keep going to the same college, and some people may never even notice that you no longer go to military classes. You can leave the academy before incurring an obligation, but doing so is a more wrenching experience, one that could embarrass your family and force you to explain your decision for the rest of your life.

Finally, if you think you are academy material, you might also qualify for entrance to an equally competitive university like Harvard, MIT, or Stanford—where the academic training will surpass that of the academy. Who ever heard, for example, of a professor at a military academy winning a Nobel Prize?

Other Things to Consider about ROTC

If you are fortunate enough to get a four-year ROTC scholarship, it can be worth as much as $30,000 toward a college education. In these days of rising tuitions and high interest rates, this can be an excellent means of paying for college.

ROTC programs are available at over a thousand institutions of higher learning, from community colleges all the way up to the Ivy League. Those schools that do not have ROTC programs often have cross-enrollment agreements with schools that do. You can pick an ROTC program at a college that matches your abilities, your geographical preferences, and your educational budget (see Appendix).

If you are not sure about military service, you can try ROTC for two years with no military obligation. If you decide to stay in you may receive a scholarship that pays for tuition, books, and uniforms and includes a tax-free allowance of $100 per month or up to $1,000 per year. Scholarship or not, you stay in the college of your choice and, for the most part, lead a normal student's life. You have the opportunities to meet people and experience things that make up a good college education.

Your school year may not differ much from that of your non-ROTC roommate, but your summers certainly will. While your roommate is goofing off or working at a boring job, you can undergo pilot training, ship out on a Navy vessel, or go to camp to practice the things you have learned in textbooks. You will have the opportunity to visit military installations and see what lies ahead.

When at last you enter the military you go in as an officer, with all the privileges and pay that come with the rank. You can travel to exciting places and face challenges that no civilian job can offer, perhaps rising up the ladder faster than your nonmilitary counterpart. After you fulfill the relatively short military obligation, you can return to civilian life or make the military your career. It's up to you.

Then there are the disadvantages. Not all of the ROTC programs will let you major in anything you want; most will restrict you to engineering or science. This is fine if it's what you planned to do anyway or if you are talented in these areas. But you should avoid getting into a subject that you may dislike.

In a related vein, college is a place to learn about yourself as well as the outside world. Many young people enter college thinking they want to do one thing only to find later on that their interests have changed. Within the first two years of ROTC (one year if you've been granted a four-year scholarship) you are free to drop out, but thereafter you are committed to military service. If your plans change, or even if your thoughts about the military change, you are in—and it's very difficult to get out.

If you flunk out of school your plans for becoming an officer go down the drain. But if you are enrolled in the final two years of ROTC you still have a military commitment. A special board reviews your case; it can discharge you or force you to enter the enlisted ranks of the service.

ROTC can be an inconvenience while you are in college. Even while not in uniform you have to keep your hair cut to military standards. On the days you take military classes you have to wear your uniform, and this may bring on good-natured or not-so-good-natured kidding, depending on the atmosphere of your college. You may not be able to participate in activities that conflict with your military ones. Finally, while your roommates are sleeping late on Saturday mornings, you may be out in the hot sun or cold wind marching and learning close-order drill.

Those summer camps or visits to military installations can throw a rock into plans for work, travel, or the chance to go to summer school. Instead of going to sea during the summer, one Naval ROTC member found himself spending a couple of weeks aboard a ship that never left the dock. He gained some useful experience, but it wasn't quite what he had in mind.

Finally, at the end of your college years your military obligation must be met. It may be three or four years, but it can seem like much longer. Suddenly you are away from civilian life while your friends are pursuing further education or starting up the job ladder. If you decide not to make a career of the military, you may feel three or four years behind everyone else when at last you return to civilian life.

Like their counterparts in the enlisted ranks, many ROTC trainees sign up without ever talking to people who are in the service and who can tell them what life in uniform is like. When you first get into ROTC you tend to think only about the benefits; the obligation doesn't confront you until it is too late.

Just like everything else in the military—for enlisted people or officers—there are some worthwhile benefits, yet they all come at a price.

ROTC Programs

The Army and Air Force each have an ROTC program, and the Navy and Marine Corps have a combined program. Descriptions of these follow.[1] The Coast Guard has no ROTC program.

ROTC programs usually have a two-year and a four-year plan. Those in the four-year plan take two years of a general military course followed by two years of a professional officer course. Those in the two-year plan take only the professional officer course. Between the sophomore and junior years everyone goes off to a field training course at a base or a ship, the four-year cadets for four weeks and the two-year cadets for a little longer. The extra time enables the two-year cadets to catch up on what they missed by not taking the general military course.

Army ROTC. This program provides college-trained officers for the U.S. Army, the Army Reserve, and the Army National Guard. It is currently offered at more than three hundred host institutions across the country with more than a thousand other schools offering Army ROTC through cross-enrollment. (For a complete list of colleges offering Army ROTC, see the Appendix.)

Army ROTC is a four-year program divided into two parts—the Basic Course and the Advanced Course. The Basic Course is normally taken in the freshman and sophomore years. Students who have participated for three years in Junior ROTC, a program conducted at twelve hundred high schools across the country, can get credit for the first year of the Basic Course. No military commitment is incurred during the first two years, when students may withdraw at any time (except

---

1. Much of the information in this section comes from *Profile*, a publication of the Department of Defense High School News Service.

for four-year scholarship holders, who incur an obligation at the beginning of their sophomore year). Subjects include management principles; national defense; military history; leadership development; and military courtesy, discipline, and customs. Uniforms, required textbooks, and materials are furnished without cost to the student.

After completing this course, selected students may enroll in the Advanced Course during the final two years of college. Instruction in this program includes further leadership development, organization and management, tactics, and administration. Cadets in the Advanced Course receive uniforms, military science textbooks, salary during an advanced camp, and a living allowance of $100 per month or up to $1,000 each year.

The six-week advanced camp is held during the summer between the junior and senior years. This camp permits cadets to put into practice the principles and theories they have learned in the classroom. Successful completion of the advanced camp is required prior to commissioning.

For community and junior college graduates and students at four-year colleges who have not taken Army ROTC during their first two years, there is a special two-year program. Students can enter this program by successfully completing a paid six-week basic camp after their sophomore year and enrolling in the Advanced Course in their junior year. Except for the basic camp, all other requirements for and obligations incurred in the two- and four-year programs are the same.

Under the Simultaneous Membership Program, a person can enlist in the Army National Guard or Army Reserve after high school, attend basic training during the summer, and enroll in the ROTC Advanced Course as early as the freshman year of college. Upon successful completion of the Advanced Course, the cadet can receive an early commission and serve as a second lieutenant with the Army National Guard or Reserve while completing the baccalaureate degree.

Army ROTC offers scholarships for four, three, and two years. Four-year scholarships are awarded on a worldwide competitive basis to U.S. citizens who will be entering college as freshmen. Three-year and two-year scholarships are awarded competitively to students who are already enrolled in college. Students who attend the basic camp of the two-year program may also compete for two-year scholarships while at camp. Two-year scholarships are available on a competitive basis for enlisted personnel on active duty in the Army. These scholarships pay for tuition, textbooks, lab fees, and other educational expenses, and they provide a living allowance of up to $1,000 each year the scholarship is in effect. The value of the scholarship depends on the tuition and other educational costs of the school attended.

Under the ROTC-Reserve Forces Duty Scholarship program, a limited number of two-year scholarships are also available to students who desire to serve in the Army National Guard or the Army Reserve in lieu of extended active duty.

Nursing students may qualify for appointment in the Army Nursing Corps through ROTC. Two-year ROTC scholarships are also available to nursing students. Recipients of these scholarships serve on active duty after appointment in the Army Nurse Corps.

Scholarship candidates are obligated to serve four years on active duty and four more in the reserve. Nonscholarship graduates may serve three years on active duty and the remainder in the reserve, or they may volunteer or be chosen to serve on Reserve Forces Duty (RFD). On RFD the active-duty obligation is from three to six months for attendance at the officer basic course, with the remaining time spent in a reserve unit.

Details about the Army ROTC program may be obtained at the ROTC detachment at participating universities or by writing to Army ROTC, Fort Monroe, Virginia 23651. Scholarship forms are available after April 1 each year from Army ROTC, P.O. Box 9000, Clifton, New Jersey 07015.

Naval ROTC. At sixty-six colleges and universities across the country, plus 114 more under the cross-enrollment system, Navy or Marine Corps commissions are granted to college students who complete either two or four years of naval science study. (For a complete list of colleges offering Naval ROTC, see the Appendix.)

The Scholarship Program offers a four-year and a two-year plan. Students compete nationally for both; qualified enlisted personnel on active duty are also eligible to apply. Those who make it are appointed midshipmen in the Naval Reserve, and are entitled to pay and benefits for the duration of the scholarship. While the scholarship is in effect the Navy pays for college tuition, fees, and textbooks, and provides uniforms and $100 cash per month.

Whereas four-year scholarship students attend military classes and drills from the time they start college, recipients of the two-year scholarships start with a six-week cram course in the summer between their sophomore and junior years. Taken at the Naval Science Institute, the course is equivalent to the freshman and sophomore Naval ROTC classes and drills.

The College Program's biggest difference from the Scholarship Program is money—you don't get much. During the first two years of the four-year College Program, members take military courses and participate in drills alongside those in the Scholarship program, but they pay their own tuition, fees, and so on. Once they enter the Advanced Course in their junior year, however, they join the Naval Reserve, whereupon the Navy provides uniforms, naval science textbooks, and $100 a month. Those in the College Program who want a better deal can apply to the Scholarship Program, where three-, two-, and one-year scholarships await.

In the two-year version of the College Program, applicants must play catch-up by attending the six-week course at the Naval Science Institute. Once they have successfully completed the course, they join the Naval Reserve and get the same benefits as those participating in the four-year version of the College Program.

Information about the College Program is available from any of the sixty-six Naval ROTC units. Students seeking scholarship information should write the Commander, Naval Recruiting Command, 4105 Wilson Boulevard, Arlington, Virginia 22203.

Air Force ROTC. This program is open to any full-time student at one of the 151 colleges and universities hosting an Air Force ROTC unit, or at one of the over six hundred institutions that have a cross-enrollment agreement with a host college. For a complete list of colleges offering Air Force ROTC, see the Appendix.

The Air Force offers scholarships annually in both the two- and the four-year plans, which are respectively called the General Military Course and the Professional Officer Course. Each scholarship pays full tuition; certain fees, including those for textbooks; and a monthly cash allowance. The scholarships are available in two-, three and a half-, and four-year allotments. Emphasis is placed on scientific or technical fields, although some scholarships are available for health professions, foreign language study, and certain nontechnical fields.

For those who would like to fly and are qualified to do so, the Air Force will pay for flying lessons, usually provided by a local civilian flying service, while they are in college. Through this program the Air Force learns how interested members are in flying and how good they are at it.

In addition to the summer field training course that everyone attends, cadets can apply for another summer training program that is held on an Air Force base. Up to eighteen hundred cadets are selected for either a two-week or a three-week session, during which they work alongside Air Force junior officers and get a chance to experience managing, leading, and working with various kinds of people. The Air Force pays their way to and from the base, and they get paid a salary while there.

A limited number of ROTC members who want to learn parachuting may volunteer for the Army's Airborne Training Program. Open to men and women, it takes three weeks and is held at Fort Benning, Georgia.

Further details on Air Force ROTC can be obtained by writing Air Force ROTC, Maxwell Air Force Base, Alabama 36112 or by contacting a ROTC unit at a local college or university.

OFFICER CANDIDATE SCHOOLS

The third way of becoming an officer in the armed forces is through officer candidate school. This is the in-house means of making the jump to the officer corps: to enter the program, you must first join the enlisted ranks. Some of the requirements are different for those who are already in the service and those who are just coming in.

Gaining admission to officer candidate school is a competitive process, and college students are advised to apply early in their senior year, for it takes several months to decide who gets in. Military personnel should make inquiries through their commanding officer.

Officer candidate school training is very intense, much like basic training. It lasts anywhere from three to five months, depending on the branch of the service. If a person continues on with flight training or any other schooling the process may last as long as a year and a half.

Graduates of officer candidate school generally incur an obligation of three or four years; those who continue with flight training usually have to stay in at least five years. Specific information on each branch is available from local recruiters.

Army Branch Immaterial Officer Candidate School. Army OCS supplies officers for the regular Army, the Army Reserve, and the Army National Guard. Officers who are commissioned through Army OCS are obligated to serve three years.

Enlisted personnel and warrant officers on active duty or in the reserves must be between nineteen and a half and twenty-nine years old on enrollment and must have completed at least two years of college class work toward a baccalaureate degree. Further information can be obtained through your commanding officer.

Civilians must be between nineteen and twenty-eight years old and must already have a baccalaureate degree. They must take a test called the Officer Selection Battery and score at least 90. Presuming they meet the qualifications, they will be paid at the rate of sergeant (E-5) during the fourteen-week training. For civilians, information on OCS can be found at an Army recruiting station.

Navy Officer Candidate School. Navy OCS applicants have to be between nineteen and twenty-seven and a half years of age at the time of commissioning. The training takes sixteen weeks at Newport, Rhode Island; graduates are commissioned as ensigns and serve in the active Navy or the Naval Reserve. The obligation for active duty is four years, except for persons going into the Supply Corps, for whom the obligation may be either three or four years.

Air Force Officer Training School (OTS). Applicants must be between twenty and a half and twenty-nine and a half at the time of application, and they must be commissioned by the age of thirty. Trainees complete a twelve-week course, and on graduation they are commissioned as second lieutenants in the Air Force. They incur an obligation of four years from the time of graduation. The training is conducted at Lackland Air Force Base near San Antonio, Texas, and graduates serve in either the active Air Force, the Air Force Reserve, or the Air National Guard.

Marine Corps Officer Candidate Class (OCC). Applicants must be between nineteen and twenty-seven and a half years of age, and they must be commissioned by age twenty-eight. The training takes ten weeks at Quantico, Virginia, and graduates are commissioned as second lieutenants. Then they spend twenty-three weeks at the Basic School, a school for officers in Quantico, Virginia, and finally leave to serve an obligation of three years in either the Marine Corps or the Marine Reserve.

The Corps also has a Marine Platoon Leader's Class (PLC), which is available to full-time qualified male freshmen, sophomores, and juniors who are attending accredited colleges. Training is conducted at Quantico, Virginia, and consists of two six-week sessions or one ten-week session during summer vacation.

Upon graduation PLC participants are commissioned as

second lieutenants—if they accept their commissions. Unlike participants in the other OCS programs, PLC members can decline to go any further at this point. If they accept the commission, they are sent to the Basic School for twenty-one weeks and then begin their military obligation of three years.

Coast Guard Officer Candidate School. College graduates or those with thirty semester hours toward a college degree can apply for a seventeen-week training session at a school in Yorktown, Virginia. Applicants must be between twenty-one and twenty-six years old. They are commissioned as ensigns and must serve at least three years.

Air National Guard Officer Training School (OTS). The Air National Guard has its own academy outside of Knoxville, Tennessee. The course lasts for six weeks and leads to a commission as a second lieutenant with a three-year obligation. In some of the fields of study a college degree is not required.

Army National Guard Officer Candidate School. Men and women in the National Guard have three options for OCS training. They can take the nine-week Reserve Component OCS, the fourteen-week active Army Branch Immaterial OCS, or a state program. The state programs usually last one year, with training taking place one weekend a month to minimize interference with civilian pursuits. Requirements vary from state to state; for information check with your local National Guard armory.

Direct Appointments

Doctors, attorneys, nurses, civil engineers, ministers, and persons in other professions needed in the military can receive direct commissions as officers. In some cases the armed forces will enlist a medical or a law student and pay his or her way through school. Direct appointments are made through highly specialized programs. More information on these can be obtained from local recruiters.

6. The Army

U.S. Army photo

THE ARMY IS THE NATION'S largest fighting force, with 777,000 men and women on active duty. Of these 109,000 are officers; 10 percent of the officers are women. Enlisted personnel total 668,000; 10 percent of these, too, are women.

Roughly two-thirds of the Army's people are stationed in the United States. The other third are scattered around the globe, as shown in the table (only countries with more than a hundred troops are included).

History

The Continental Army was established in 1775 with George Washington in command. After the revolutionary war the country turned to other matters, and when the War of 1812 came along the Army was ill prepared. The development of the U.S. Military Academy at West Point and greater attention from Congress fostered an Army that handled itself well in the Mexican War of the 1840s.

Location	*Personnel*
UNITED STATES	
Continental United States	475,100
Alaska	7,700
Hawaii	17,900
WESTERN AND SOUTHERN EUROPE	
Belgium	1,300
Germany	209,800
Greece	600
Italy	4,200
Netherlands	900
Turkey	1,200
United Kingdom	200
EAST ASIA AND THE PACIFIC	
Japan and Okinawa	2,200
Johnston Atoll	140
Philippines	100
Korea	32,300
MIDDLE EAST	
Egypt	1,300
Saudi Arabia	170
CENTRAL AMERICA AND CARIBBEAN	
Honduras	1,400
Panama	7,000
Puerto Rico	500

The Civil War, in which the officers of the Army led troops against each other, proved what a deadly force the Army had become. The strategy and tactics perfected in the 1860s conflict, particularly the use of the railroads and the telegraph, served as a pattern for future wars around the world.

The needs of World War I swelled the Army to over 4 million men, and the height of World War II saw over 11 million men in uniform.

The Army fought communist forces for the first time in the Korean Conflict, eventually securing the border of what is now South Korea.

The 1960s saw the Army in Vietnam, in a conflict that bitterly divided America and became one of the greatest frustrations in the history of the Army. North Vietnamese forces proved elusive in the jungles of Vietnam despite large numbers of U.S. troops and the use of ultramodern combat hardware.

Since Vietnam the Army has undergone a period of self-examination, and changes have been made. The biggest has been the end of the draft, in 1973. The switch to all-volunteer forces has brought many new features to the Army. Vocational and educational opportunities have been emphasized, and more opportunities have opened for women.

Length of Hitch

Officers. All commissioned officers have to serve in the Army for eight years, splitting their obligation between active and reserve duty. It breaks down like this, in years:

	Active Duty	*Ready Reserve*	*Standby Reserve**
ACTIVE ARMY			
U.S. Military Academy	5	0	3
ROTC with scholarship	4	0	4
ROTC without scholarship	3	0	5
Officer candidate school	3	0	5
ARMY RESERVE			
Reserve Forces Duty	¼–½	remainder of 8	0

*Officers can stay in the Ready Reserve if they prefer.

The time you spend on active duty pays the Army for your schooling. If you want to take further training the Army is willing—but you have to stay longer.

Here are some of the things that can add to your obligation:

☐ *Graduate school.* If the Army sends you to graduate school you must stay in uniform two years for every one year in school.
☐ *Promotion.* If you accept a promotion you obligate yourself for another two years.
☐ *Aviation training.* Although the Army is not the Air Force, it has over eighty-five hundred aircraft, mostly helicopters. Flight training for helicopter duty lasts approximately nine months, and aviators must serve four years after completing flight school.

An important note: An officer's extra obligations can be served simultaneously. This means that, if you go through Army ROTC and then take flight training, you will not owe the Army eight years. If you entered flight training right after commissioning, you would probably be obligated to serve only five years.

Enlisted. Those who enlist in the Army commit themselves to eight years of service in a combination of active and reserve duty. Those who join the Army Reserve spend only a few months on active duty before going into Ready Reserve.

The Army offers tours of two to four years active duty. The two-year hitch is the shortest of all the services.

Here is a breakdown of the enlistment programs by years:

	Active Duty	*Ready Reserve*	*Standby Reserve*
Regular Army	2	0	6
	3	0	5
	4	0	4
Army Reserve	⅓–⅔	remainder of 8	0

Basic Training

"Going through basic" is perhaps the most feared and the most bragged-about experience in the Army this side of combat. Forty to fifty recruits form a platoon and are trained by a non-commissioned officer known as the drill instructor.

Despite electronic vision enhancers, personnel sensors, and other high-tech hardware, the Army is first and foremost a ground force. When you take basic training you can expect around a hundred miles of marching—day and night—camping, camouflage, and plenty of time on the rifle range. It all lasts for seven weeks.

A typical day. Here is an hour-by-hour schedule for a day in the sixth week of basic training. This is perhaps the toughest day you'll have, for instead of hitting the sack at night, you'll hit the road for a night march and bivouac.

4:30 A.M.	First call
4:45–5:30	Physical training
5:30–6:00	(Decided by drill instructor)
6:00–6:30	Breakfast
6:30–7:00	Inspection
7:00–9:50	U.S. weapons familiarization
9:50–12:00 P.M.	Concealment and maneuvers
12:00–1:00	Lunch
1:00–4:00	Hand grenade assault course
4:00–5:00	Dinner
5:00–7:00	Personal time
7:00–Completion	Evening activities
12:00 A.M.	March and bivouac

In the training sessions you will learn how the Army operates and how you are expected to operate in combat. Topics covered include such things as:

First aid	Military customs and courtesies
M-16 familiarization	Camouflage
Grenade training	Gas mask training
Close-order drill	Hygiene and sanitation
Promotion	Uniform Code of Military Justice

Location. Male recruits undergo basic training at all of the following bases. Women train only at those marked with asterisks.

Installation	*Nearest City*
Fort Benning	Columbus, Georgia
Fort Bliss	El Paso, Texas
Fort Dix*	Trenton, New Jersey
Fort Jackson*	Columbia, South Carolina
Fort Knox	Louisville, Kentucky
Fort Leonard Wood	Waynesville, Missouri
Fort McClellan*	Anniston, Alabama
Fort Sill	Lawton, Oklahoma

Jobs

The Army offers twenty-eight categories of jobs, which it calls Military Occupational Specialties, or MOS's for short. There are about three hundred entry-level MOS's, ranging from small weapons repairer to club manager, from radio and TV broadcasting production coordinator to tank turret repairer. Some MOS's, particularly those that require a lengthy training period, are not available for short enlistments. Special programs, such as the Army College Fund, have narrower lists of MOS's—usually combat-related ones—from which to choose.

Women are barred from combat jobs such as driving tanks, although they can serve in combat support as truck drivers, medics, and so on. Some MOS's that may not seem to be combat-related are also barred to women. Chief among these are the building trades, such as plumbing, interior electrical work, and the operation of heavy construction equipment. The Army now tests women's physical abilities before recommending jobs that require great muscular strength.

Job training, called Advanced Individual Training to distinguish it from Basic Training, can last anywhere from a few weeks to almost a year. Here are three job categories and a sampling of the work done in each.

☐ *Air defense artillery.* A person in this MOS area will emplace, assemble, test, maintain, and fire air defense weapons systems. Related tasks include operating fire control equipment, radar, computers, automatic data transmission systems, and associated power-supply equipment. The training focuses on acquiring basic mechanical and electrical knowledge.

This MOS area includes combat jobs that are barred to women.

☐ *Administration.* Persons in this field maintain pay records of military personnel, prepare vouchers for payment, prepare reports, and disburse funds. They perform tasks such as budgeting, allocating, auditing, and compiling and analyzing statistical data.

☐ *Transportation.* Those in this MOS area drive and maintain passenger vehicles and light, medium, and heavy cargo vehicles. They may pilot small boats or even perform as air traffic controllers.

Army Apprenticeship Program

The Army Apprenticeship Program enables you to earn a U.S. Department of Labor certificate that is recognized by civilian employers. It states that you have achieved journeyman status and are skilled at your particular trade.

After you receive the training you log your work hours in a special book. Depending on the trade you choose, you have to log between two thousand and eight thousand hours to get your certificate. You can do so, however, only if the Army assigns you to work in your chosen field, which it tries to do. But sometimes the Army needs you elsewhere, and you will not have time during your enlistment to complete the log. You can always reenlist, but that's another story.

Bonuses

Since Army recruits have their choice of training, sometimes there is a shortage of people in a particular job. Often those jobs are combat-related and have little equivalent in civilian life. Other times the jobs require extensive schooling. To fill these jobs the Army offers cash, payable when you complete the necessary training. Once the MOS attracts enough people the offer is withdrawn.

For this reason it is impossible to list the jobs that rate bonuses right now. For such a list you need to check with your Army recruiter. Jobs for which bonuses have been given in the past include:

Electronic warfare/signal intelligence voice interceptor	$8,000
Infantryman	5,000
Wire system installer	2,500

Job List

Following is an abbreviated list of Army MOS's. Primarily entry-level positions, they are grouped under headings defined by the Army. The occasional words in capital letters refer to specific weapon systems. This list comes from the 1986 U.S. Army Pocket Recruiting Guide, but to get the most current job information you should check with a recruiter.

The symbols after some MOS's indicate special benefits or limitations that apply to them, as listed here.

2-year	Two-year enlistment option
ACF	Army College Fund (see Chapter 15)
EB$	Enlistment bonus
AAP	Army Apprenticeship Program
*	Closed to women

Note: Some jobs qualify for both the Army College Fund and an enlistment bonus. You can choose one or the other, but not both.

Administration

Secretary
Legal clerk
Court reporter
Administrative specialist (2-year, ACF)
Chapel activities specialist (ACF)
Finance specialist
Accounting specialist
Personnel administration specialist
Personnel management specialist

Personnel records specialist
Personnel actions specialist
Equal opportunity NCO
Physical activities specialist

Air defense artillery (ADA)

Light air defense artillery crewmember (Reserve)*
ADA operations and intelligence assistant (2-year, ACF)
Defense acquisition radar operator (ACF)*
ADA short range missile crewman (2-year, ACF, EB$5000)*
ADA short range gunnery crewman (2-year, ACF, EB$6000)*
Man Portable Air Defense System crewman (2-year, EB$6000, ACF)*
PATRIOT missile crewmember (EB$6000)
Air defense missile crewmember (2-year, ACF, EB$6000)

Air defense missile maintenance

NIKE test equipment repairer (AAP)
NIKE-HERCULES missile launcher repairer (AAP)
NIKE track radar repairer (AAP)
NIKE high power radar simulator repairer (AAP)
Improved HAWK firing section mechanic (AAP)
Improved HAWK fire control mechanic (AAP)
Improved HAWK information coordination central mechanic (AAP)
Improved HAWK fire control repairer (AAP)
Improved HAWK pulse radar repairer (AAP)
Improved HAWK continuous wave radar repairer (AAP)
Improved HAWK launcher/mechanical systems repairer (AAP)
Defense acquisition radar mechanic (AAP)
NIKE-HERCULES fire control mechanic (AAP)
Improved HAWK master mechanic
PATRIOT systems mechanic
HERCULES electronics mechanic (AAP)
AN/TSQ-73 air defense artillery command and control system operator/repairer (AAP)
Air defense radar repairer (AAP)

Ammunition

Nuclear weapons electronics specialist (AAP)
Ammunition specialist (ACF, 2-year, EB$3000)
Explosive ordinance disposal specialist (ACF, EB$4500, 2-year)
Nuclear weapons maintenance specialist (AAP)
Ammunition stock control and accounting specialist

Armor

Cavalry scout (2-year, ACF, EB$4000)*
M48-M60 armor crewmember (2-year, ACF, EB$6000)*
M1 armor crewman (2-year, ACF, EB$7000)*

Automatic data processing

Punchcard machine repairer (AAP)
Decentralized automatic service (DAS) system computer repairer (AAP)
NCR 500 computer repairer (AAP)
DSTE repairer (AAP)
ADMSE repairer

UNIVAC 1004/1005 DCT 9000 system repairer (AAP)
IBM 360 repairer (AAP)
Field artillery computer repairer
Card and tape writer (Reserve)
Computer/machine operator
Programmer/analyst

Aviation communications–electronics systems maintenance

Ground control approach radar repairer (AAP)
Aerial radar sensor repairer
Aerial electronic warning defense equipment repairer
Aerial surveillance radar repairer (Reserve)
Aerial surveillance infrared repairer (Reserve)
Avionic mechanic (AAP)
Avionic communications equipment repairer (AAP)
Avionic navigation and flight control equipment repairer (AAP)
Avionic equipment maintenance supervisor
Avionic special equipment repairer (AAP)
Aerial surveillance photographic equipment repairer (Reserve)

Aviation maintenance

Airplane repairer (AAP)
Observation airplane repairer
Utility helicopter repairer (AAP)
Tactical transport helicopter repairer (AAP)
Medium helicopter repairer (AAP)
Observation scout helicopter repairer (AAP)
Heavy lift helicopter repairer (AAP)
Attack helicopter repairer (AAP)
Aircraft powerplant repairer (AAP)
Aircraft powertrain repairer
Aircraft electrician (AAP)
Aircraft structural repairer (AAP)
Aircraft pneudraulics repairer (AAP)
Aircraft fire control repairer (AAP)
Aircraft components repair supervisor
Aircraft weapon systems repairer (AAP)

Ballistic land combat missiles and light air defense weapons systems maintenance

PERSHING electronics materiel specialist (ACF)
PERSHING electronics repairer (AAP)
VULCAN system mechanic (AAP)*
CHAPARRAL system mechanic (AAP)
Land combat support system test specialist/LANCE repairer
TOW/DRAGON repairer (AAP)
VULCAN repairer (AAP)
CHAPARRAL REDEYE repairer (AAP)
SHILLELAGH repairer (AAP)
Forward area alerting repairer
PERSHING electrical-mechanical repairer (AAP)

Band

Cornet or trumpet player
Baritone or euphonium player
French horn player
Trombone player (ACF)
Tuba player
Flute or piccolo player
Oboe player (ACF)
Clarinet player (ACF)
Bassoon player (ACF)
Saxophone player
Percussion player
Piano player
Brass group leader
Woodwind group leader

Percussion group leader
Special bandperson
Guitar player

Chemical

Smoke operations specialist (2-year, ACF, EB$2500)*
NBC specialist (ACF, EB$2500, 2-year)
Chemical laboratory specialist

Combat engineering

Combat engineer (2-year, ACF, EB$3500, AAP)*
Bridge crewmember (2-year, ACF, EB$4500, AAP)*
Atomic demolition munitions specialist*
Engineer track vehicle crewman (2-year, ACF, EB$2000)*

Communications-electronics maintenance

Weapons support radar repairer (AAP)
Combat area surveillance radar repairer (AAP)
Tactical microwave systems repairer (AAP)
Strategic microwave systems repairer (AAP)
Satellite communications (STRACOM) equipment repairer (AAP, ACF)
Field radio repairer (AAP)
Teletypewriter repairer (AAP)
Field general COMSEC repairer (AAP)
Field systems COMSEC repairer (AAP)
Fixed ciphony repairer (AAP)
Fixed cryptographic equipment repairer (AAP)
Fixed station radio repairer (AAP)
Special electronic device repairer (AAP)
Calibration specialist (AAP)
Dial/manual central office repairer (AAP)
Electronic switching repairer (AAP)

Communications-electronics operations

Tactical satellite/microwave systems operator (AAP)
Strategic microwave system operator (AAP)
Single channel radio operator (2-year, ACF, EB$4000)
Combat signaler (2-year, ACF, EB$4000)
Multichannel communications equipment operator (2-year, ACF, AAP)
Tactical circuit controller
Tactical communications systems operator/mechanic (AAP, 2-year, ACF, EB$3000)
Station technical controller (EB$2000)
Wire system installer (AAP, EB$2500, ACF)
Antenna installer specialist (AAP)
Cable splicer (AAP)
Wire systems operator (2-year, ACF, EB$3500)
Combat telecommunications center operator (2-year, ACF, AAP)
Automatic data telecommunications operator (2-year, ACF)

Electronic warfare/cryptological operations

Electronic warfare/signal intelligence identifier/locator
Signal security specialist
Electronic warfare/signal intelligence Morse interceptor (ACF, EB$8000)

Electronic warfare/signal intelligence non-Morse interceptor (ACF, EB$5000, 2-year)
Electronic warfare/signal intelligence analyst (ACF, [Language only: EB$7000])
Electronic warfare/signal intelligence voice interceptor (ACF, EB$8000)
Electronic warfare/signal intelligence noncommunications interceptor

Electronic warfare/intercept systems maintenance

Field artillery

Cannon crewman (2-year, ACF, EB$7000)*
TACFIRE operations specialist (ACF, EB$2000)*
Cannon fire direction specialist (2-year, ACF, EB$3500)*
Fire support specialist (2-year, ACF, EB$3500)*
Multiple launch rocket system (MLRS) crewmember (EB$5000, 2-year, ACF)*
Field artillery firefinder radar operator (ACF, EB$2000)*
LANCE crewmember/MLRS sergeant (2-year, ACF, EB$5000)
PERSHING missile crewmember (2-year, ACF, EB$5000)
MLRS/LANCE operation fire direction specialist (ACF)*
Field artillery radar crewmember*
Field artillery surveyor (2-year, ACF, EB$2500, AAP)*
Field artillery meteorological crewmember

Food service

Food service specialist (2-year, ACF, AAP, EB$4000)
Hospital food service specialist (AAP)

General engineering

Carpentry and masonry specialist (AAP)
Structures specialist (AAP)
Materials quality specialist
Plumber*
Firefighter (AAP)
Water treatment specialist (AAP)
Interior electrician (AAP)*
Prime power production specialist (AAP)
Transmission and distribution specialist*
Industrial gas production specialist (Reserve)
Heavy construction equipment operator (AAP, EB$2500)
Lifting and loading equipment operator (AAP)
Quarrying specialist (AAP)
Concrete and asphalt equipment operator (AAP)
General construction equipment operator (2-year, ACF, EB$2000)
Technical drafting specialist (AAP)
Construction surveyor (ACF, AAP)

Infantry

Infantryman (2-year, ACF, EB$5000)*

Law enforcement

Military police
Correctional specialist
Special agent

Mechanical maintenance

Fire control instrument repairer
Office machine repairer
Metal worker (AAP)
Machinist (AAP)

Small arms repairer (AAP)
Self-propelled field artillery turret mechanic (ACF, EB$3500)*
M-1 Abrams tank turret mechanic (2-year, ACF, EB$4500)*
Fire control systems repairer
Tank turret repairer (2-year, ACF, AAP, EB$2500)
Artillery repairer (AAP, 2-year, ACF, EB$2500)
M60A1/A3 tank turret mechanic (ACF, EB$3500, AAP, 2-year)*
M2/Bradley fighting vehicle systems mechanic (2-year, ACF, EB$3500)*
Utilities equipment repairer (AAP, EB$2500)
Power generation equipment repairer (2-year, ACF, EB$4000)
Construction equipment repairer (AAP, 2-year, ACF, EB$3000)
Light wheel vehicle power generation repairer (ACF, AAP, EB$3500, 2-year)
Self-propelled field artillery system mechanic (AAP, ACF, EB$3500, 2-year)*
M-1 Abrams tank system mechanic (AAP, ACF, EB$3500, 2-year)*
Fuel and electrical systems repairer (ACF, EB$4000, 2-year)
Wheeled vehicle/track vehicle repairer (AAP, ACF, EB$4000, 2-year)
Quartermaster and chemical equipment repairer (AAP, ACF, EB$3500, 2-year)
M60A1/A3 tank systems mechanic (EB$3500, AAP, ACF, 2-year)
Heavy wheel vehicle mechanic (AAP, ACF, EB$3500, 2-year)
Improved TOW vehicle/infantry fighting vehicle/cavalry fighting vehicle system mechanic (ACF, EB$3500, AAP, 2-year)*
Wheel vehicle repairer (AAP, ACF, EB$3500, 2-year)
Track vehicle repairer (AAP, ACF, EB$3500)

Medical

Biomedical equipment specialist, basic
Orthodontic specialist
Dental laboratory specialist
Optical laboratory specialist
Patient administration specialist
Medical specialist (ACF, 2-year)
Practical nurse (ACF, EB$5000, 2-year)
Operating room specialist
Dental specialist
Psychiatric specialist
Behavioral science specialist
Orthopedic specialist
Physical therapy specialist
Occupational therapy specialist
Cardiac specialist
X-ray specialist
Pharmacy specialist
Veterinary food inspection specialist
Environmental health specialist
Animal care specialist
ENT specialist
Respiratory specialist (EB$5000)
Eye specialist
Medical laboratory specialist
Biological science assistant

Military intelligence

Intelligence analyst
Image interpreter (AAP, ACF)
Aerial sensor specialist (OV-1D)
Ground surveillance systems operator*
Counterintelligence agent (ACF, EB$2500)
Area intelligence specialist

Petroleum

Petroleum supply specialist (AAP, ACF, EB$4500, 2-year)
Petroleum laboratory specialist

Public affairs and audio-visual

Radio/television systems specialist (AAP)
Audio-visual equipment repairer (AAP)
Journalist
Broadcast journalist
Illustrator (AAP)
Still photographic specialist (AAP)
Motion picture specialist (AAP)
Audio/television specialist (AAP)

Special forces

Supply and service

Parachute rigger
Fabric repair specialist (EB$2000)
Laundry and bath specialist (EB$2500)
Graves registration specialist
Equipment records and parts specialist (2-year, ACF, EB$4000)
Medical supply specialist
Materiel control and accounting specialist (EB$2000, ACF)
Materiel storage and handling specialist (EB$2500, 2-year, ACF)
Subsistence supply specialist (2-year, ACF, EB$2000)
Unit supply specialist (2-year, ACF)

Topographic engineering

Topographic instrument repair specialist
Cartographer
Topographic surveyor
Photo and layout specialist
Photolithographer

Transportation

Cargo specialist (2-year, ACF, EB$3500)
Watercraft operator
Watercraft engineer (AAP)
Marine hull repairer (AAP)
Motor transport operator (2-year, ACF, EB$4000)
Locomotive repairer (Reserve)
Railway car repairer (Reserve)
Airbrake repairer (Reserve)
Locomotive electrician (Reserve)
Railway section repairer (Reserve)
Locomotive operator (Reserve)
Train crewmember (Reserve)
Railway movement coordinator (Reserve)
Traffic management coordinator
Meteorological observer
Air traffic control (ATC) tower operator
ATC radar controller

7. The Navy

Official U.S. Navy photograph

For over two hundred years the Navy has been the chief means of projecting U.S. military might around the world. From the wooden ships of colonial times to the present-day Trident submarines, the U.S. Navy has accumulated a record of victories unmatched by any other country.

Navy sailors go out to sea in 556 ships and more than six thousand aircraft. This fighting force is staffed by a total of 569,000 men and women on active duty. Of the 71,000 officers, 10 percent are women. Of the 499,000 enlisted personnel, 9 percent are women.

At any given time, of course, a great many Navy people are on ships; others are scattered among bases around the world. (Those en route between bases or in locations with fewer than 100 sailors are excluded from the table.)

History

The Navy was officially established in 1798, but its celebrated birthday is October 13, 1775, when the Second Continental Congress authorized the purchase of two vessels. That fledgling Navy is best remembered by a quote from John Paul Jones, captain of the *Bonhomme Richard.* In the thick of a battle whose outcome was unclear, he was asked if he was ready to give up. Jones retorted, "I have not yet begun to fight!" This spirit set the tone for future naval battles.

The Navy protected the young United States, and most Navy battles were fought to keep the vital sea lanes open. In its most far reaching move, the Navy carried out a daring raid on a group of Barbary pirates in the Mediterranean Sea.

The Civil War brought forth another famous Navy quote, this time from David Farragut as he told his crew to disregard the mines in the Mobile harbor. "Damn the torpedoes! Full steam ahead!" he thundered. Farragut was not a man to waste words.

Wooden ships gave way to steel ones, and in 1898 the United States surprised Spain and won a victory under Admiral Dewey in the Philippine Islands.

In World War I the Navy ferried millions of men and tons of materiel to Europe and for the first time waged ex-

Location	Personnel
UNITED STATES	
Continental United States	271,900
Alaska	2,000
Hawaii	12,800
Afloat	178,400
WESTERN AND SOUTHERN EUROPE	
Belgium	100
Germany	300
Greece	500
Iceland	1,700
Italy	4,900
Portugal	400
Spain	3,800
Turkey	100
United Kingdom	2,400
Afloat	20,900
EAST ASIA AND THE PACIFIC	
Australia	400
Guam	4,600
Japan and Okinawa	7,300
Philippines	6,000
Korea	400
Afloat	17,600
AFRICA, MIDDLE EAST, AND SOUTH ASIA	
Diego Garcia*	1,200
Afloat	10,500
WESTERN HEMISPHERE OUTSIDE THE UNITED STATES	
Bermuda	1,400
Canada	400
Cuba (Guantanamo Bay)	1,900
Panama	500
Puerto Rico	3,000
Afloat	2,000
ANARCTICA	100

*An island in the Indian Ocean.

tensive antisubmarine warfare. World War II began for the United States with a devastating attack by the Japanese on Pearl Harbor, Hawaii. Much of the Pacific fleet was sunk in that bold move, and a Japanese admiral was said to have remarked, "I fear we have awakened a sleeping giant." His words proved prophetic.

Although the Korean and Vietnamese conflicts included

none of the epic sea battles of World War II, the Navy used battleships to bombard positions far inland, and Navy fliers flew extensive missions over battle areas. In Vietnam the Navy patrolled the rivers and supported the Marines and the Army on land.

In recent years the Navy has concentrated on modernizing the fleet and building a new generation of submarines. The missiles carried by these silent ships constitute one-third of the nation's nuclear defense system.

Length of Hitch

Officers. All commissioned officers are obligated to serve in the Navy for eight years—four or five years on active duty, and the remainder in either the Ready Reserve or the Standby Reserve.

The obligation breaks down like this, in years:

	Active Duty	*Ready or Standby Reserve*
U.S. Naval Academy	5	3
Naval ROTC	4	4
Officer candidate school	4	4

Your four or five years of active service more or less pay back the Navy for the schooling you have received. If you take additional training—if the Navy invests even more in you—then your obligation increases accordingly.

Here are some things that can add to your obligation:

☐ *Graduate school.* If the Navy pays for your graduate schooling you must stay in uniform one year for every year of school you attend. If the Navy sends you to medical school the same rule applies, but you begin working off the obligation after internship and residency.

☐ *Promotion.* Getting promoted during your initial obligation period does not mean you'll have to serve longer. Once you reach the rank of lieutenant commander, however, you owe the Navy three years whenever you accept a higher rank.

□ *Aviation training.* Flight training is perhaps the most expensive schooling in the Navy, and for that reason the obligation is the longest. If you become a flight officer, such as a navigator, bombardier, or an electronics officer, you must serve four years after completing flight school. The training averages a year in length.

If you decide to become a Navy pilot, then you can add six and a half years to your obligation. Pilot training takes one and a half years, and then you owe the Navy five more years.

An important note: Not all of these add-ons are cumulative. For instance, if you were to go through Naval ROTC and then take pilot training you would not owe the Navy ten and a half years. If you were to go into flight training right after you were commissioned you would have to serve the Navy only six and a half years.

Enlisted. Everyone who enlists in the Navy signs up for an eight-year hitch. How that eight years is spent, however, depends on the enlistment plan you choose.

If you join the Naval Reserve, you need spend only a minimal amount of time on active duty—usually just enough to cover basic training and job training.

Those who opt for the active Navy must spend three to six years on active duty, and the remainder of the eight years in the reserve. If you spend eight years on active duty you have no obligation to serve in the reserve.

As with officers, the amount of training an enlisted person takes may affect the length of active duty. For example, if you were to select the Advanced Electronics Field Training you would have to sign up for six years.

Why? In advanced electronics or a similar field, your training will take almost two years to complete. If you signed up for only three years then the Navy would get only one year of work from you before you would be eligible to leave. As elsewhere in the military, the more the Navy invests in you the longer they expect you to stay.

Here are the options, in years:

	Active Duty	*Ready Reserve*	*Standby Reserve**
ACTIVE NAVY	3	2	3
	4	0	4
	5	0	3
	6	0	2
NAVAL RESERVE			
Active Mariner Program	3	2	3
Active Mariner Apprenticeship Training Program	3	2	3
Training and Administration of the Reserves Program	4	0	4
Ready Mariner Program	½–1½	remainder of 8	0

*You can serve in the Ready Reserve instead, if you prefer.

Basic Training

Groups of approximately eighty recruits make up a company, which is under the watchful eye of the company commander, the equivalent of an Army drill instructor. He or she leads the company through the seven-week training period.

One of the first things a recruit learns is the way the Navy talks. The Navy is very proud of its seafaring traditions, and many of the terms used on land and water come from the days of wooden ships. A bathroom is called a head. The hospital or clinic is called a sick bay, and the kitchen is a galley. The drinking fountain is a scuttlebutt, and candy or gum is called gedunk.

A typical day. Basic training in the Navy incorporates intensive mental and physical activities. Your day might shape up like this.

4:30 A.M.	Reveille
5:10–5:50	Physical training

6:00–7:20	Barracks cleanup and breakfast
7:30–8:10	(Decided by company commander)
8:20–9:00	Training period 1
9:10–9:50	Training period 2
10:00–10:40	Training period 3
10:50–11:30	Training period 4
11:40–12:20 P.M.	Training period 5[1]
12:30–1:10	Training period 6[1]
1:20–2:00	Training period 7
2:10–2:50	Training period 8
3:00–3:40	Training period 9
3:50–4:30	Training period 10
4:40–5:20	Training period 11
5:30–6:00	Evening meal
6:00–7:30	Shower, shave, and shine shoes
7:30–8:15	All hands on cleaning stations; clean barracks
8:15–9:25	Recruit instructions—term, rank, knot of the day, etc.—and night bunk check
9:25–9:30	Preparation for taps
9:30	Taps

The training periods will find you learning about life on ship and such things as Navy promotions and policies and military drill. Other topics include:

Accident prevention
Chain of command
Damage control
Hand salutes and greetings
Navy mission and organization
Personal hygiene
Watchstanding
Basic deck seamanship
Cultural adjustments
Firefighting
History of the Navy
Officer recognition
Survival at sea
Uniform Code of Military Justice

1. One of these is the lunch period.

The physical part. Physical training is held every day, and there are even remedial physical workouts for people who are having a difficult time. A recruit is tested four times during basic training; here are some of the requirements for the final men's test:

Exercise	*Amount Required*
Pushups (hands under chin)	20
Situps (six-count)	15
Jumping jacks	75
Running	2¼ mi. in 18 min.

As might be expected, the Navy teaches its people to swim. You must be able to—

☐ Enter the water feet first from a height of five feet and float or tread water for five minutes;
☐ Enter the water in the deep end of a pool and swim fifty yards, using any stroke, keeping your head above the water; and
☐ Put on and care for the inherently buoyant and the CO_2-inflatable life jackets.

Location. Men undergo basic training at Great Lakes, Illinois; Orlando, Florida; or San Diego, California. Women undergo basic only in Orlando.

Jobs

The Navy offers approximately twenty-four job categories, ranging from chaplain's assistant to aviation antisubmarine warfare operator. Jobs are called *ratings* in the Navy. As in most of the other branches of the service, to be eligible for the more advanced ratings you have to plan to stay in longer.

The needs of the Navy and your ASVAB scores determine which job ratings you qualify for. Those two, plus the amount of time you decide to stay, will determine the ratings from which you can choose.

It's a rare person in the Navy who never serves on a ship, and this poses problems for women, who are forbidden by law from serving on combat vessels. They can ship out for brief periods when a vessel does not have a combat assignment, and they can serve anytime on auxiliary or noncombat ships, but the combat restrictions severely limit the number of jobs available to women. The Department of Defense estimates that 14 percent of the Navy ratings are barred to women.

Once a rating is selected, the Navy guarantees the recruit the training that he or she signs up for. After basic training the sailor goes to "A" school to learn the skill. "A" schools are all over the country—some quite far from the sea—and the length of schooling ranges from six weeks to more than forty weeks.

Here are three ratings and the length of training required:

☐ *Boatswain's mate.* Another name for this rating might be *general sailor.* All ships need people who can work on the deck, performing rigging, maintenance, and repair. The work is often outside, and it requires a lot of physical strength. "A" school training takes from six to twelve weeks, but most of the knowledge in this rating comes on the job.

☐ *Data systems technician.* With this rating you would maintain, adjust, and repair digital computers, video processors, tape units, and computer equipment. Persons seeking this rating study basic electricity and electronics for eight weeks at Great Lakes, Illinois, then go to San Francisco for twenty-two weeks of "A" school, where they study basic computer maintenance and theory. They can continue in further training if desired.

☐ *Air traffic controller.* Navy air traffic controllers are responsible for the safe, orderly, and speedy movement of aircraft in and out of landing areas. You may work on land or on the deck of an aircraft carrier. The "A" school is located in Memphis, Tennessee, and lasts for thirteen to fourteen weeks.

Advanced Training

For those who are willing to enlist for six years, the doors to advanced training are open. The schooling alone may take two years. In these programs the applicants start out two pay grades ahead of the regular enlistees and receive an E-4 grade—petty officer third class—upon graduation. Here is a brief description of three advanced programs.

☐ *Nuclear Field Program.* Following basic and "A" school, you go to the Nuclear Power School in Orlando for twenty-four weeks. Then you are assigned to a six-month on-the-job training program at a land-based nuclear power plant. Once this is completed you are ready for assignment to a surface ship, or, if accepted, to a submarine.

☐ *Advanced Electronics Field Program.* First you go through basic, then you go through twenty-two to thirty-six weeks of "A" school, where you study the basics of electronics systems, magnetic amplifiers, and remote control systems. Once this material is mastered you enroll in more advanced schools and learn about the Navy's sophisticated electronic gear. After that you will probably ship out and work with data systems, communications equipment, or sonar technology.

☐ *Advanced Technical Field Program.* As with the other advanced programs you go through basic, then "A" school, and then more specialized training. In each session you can choose from varied skills to learn.

The following are open to all qualified Navy personnel:

1. *Hospital corpsman:* one of eighteen different specialties among medical technicians.
2. *Radio technician:* operates and maintains submarine radio telegraph and radio teletype machines.
3. *Hull maintenance technician:* performs plate or high-pressure pipe welding, or heat treatment and nondestructive testing of metals.

4. *Interior communications electrician:* submarine communications electrician, automatic telephone and gyrocompass technician, or closed-circuit TV maintenance technician.

The following are open to men only:

1. *Gas turbine systems technician:* maintains and operates gas turbine engines and associated machinery aboard the newer ships in the fleet.
2. *Boiler technician:* operates shipboard equipment that produces steam for propulsion engines and the generation of electric power.

Bonuses

Like the other branches of the service, the Navy occasionally has trouble filling certain positions. Perhaps the working conditions are less than desirable, as they would be in an engine room that is hot, noisy, and below the waterline of a ship. Or perhaps very specialized training is required. Language schools, especially for obscure tongues such as Farsi, may offer bonuses to attract recruits.

The ratings that have bonuses attached change from time to time as the needs of the Navy change. Here are some ratings for which the Navy has paid bonuses in the past, as well as the amount of these bonuses. To get an up-to-date list, check with your Navy recruiter.

Nuclear fields	$3,750–6,000
Boiler technician	1,500

Job List

Following is a list of most of the entry-level jobs in the Navy. Unlike the Army and the Air Force, the Navy tends to give only a general title for a job that actually has several subcategories. For example, as an aviation machinist's mate you might specialize in maintaining and repairing helicopter engines, jet engines, or engines on fixed-wing aircraft. Because of this practice, the Navy appears to have fewer kinds of jobs than it actually does.

The ratings that are barred to women are marked with asterisks. Women should keep in mind, in addition, that by law they cannot serve on combat vessels, so even if a rating is open to them, positions within it may be unavailable because they are on warships.

Administration

Legalman
Navy counselor
Personnelman
Postal clerk
Yeoman
Religious program specialist

Air traffic control

Air traffic controller

Aviation ground support

Aviation boatswain's mate (launching and recovery equipment)*
Aviation boatswain's mate (fuels)*
Aviation boatswain's mate (aircraft handling)*
Aviation support equipment technician
Aviation support equipment technician (electrical)
Aviation support equipment technician (mechanical)

Aviation maintenance/weapons

Aircrew survival equipmentman
Aviation antisubmarine warfare technician
Aviation electrician's mate
Aviation electronics technician
Aviation fire control technician*
Aviation machinist's mate
Aviation maintenance administrationman
Aviation ordnanceman
Aviation structural mechanic (safety equipment)
Aviation structural mechanic (hydraulics)
Aviation structural mechanic (structures)

Aviation sensor operations

Aviation antisubmarine warfare operator (acoustic)*
Aviation antisubmarine warfare operator (helicopter)*
Aviation antisubmarine warfare operator (non-acoustic)*

Communications

Radioman

Construction

Builder
Construction electrician
Construction mechanic
Engineering aid
Equipment operator
Steelworker
Utilitiesman

Cryptology

Cryptologic technician

Data systems

Data processing technician
Data systems technician

General seamanship

Boatswain's mate
Signalman

Health care

Dental technician (general)
Dental technician (prosthodontics)
Dental technician (repair)
Hospital corpsman

Intelligence

Intelligence specialist

Logistics

Aviation storekeeper
Disbursing clerk
Mess management specialist
Ship's serviceman
Storekeeper

Marine engineering

Boiler technician*
Electrician's mate
Engineman
Gas turbine system technician (electrical)*
Gas turbine system technician (mechanical)*
Interior communications electrician
Machinist's mate*

Master-at-arms

Media

Illustrator draftsman
Journalist
Lithographer
Photographer's mate

Meteorology and oceanography

Aerographer's mate

Musician

Ordnance systems

Gunner's mate*
Gunner's mate (guns)*
Gunner's mate (missiles)*
Mineman
Missile technician*
Torpedoman's mate (submarine)
Torpedoman's mate (surface)
Torpedoman's mate (technician)
Weapons technician

Sensor operations

Electronics warfare technician*
Ocean systems technician (analyst)
Ocean systems technician (maintainer)
Sonar technician (surface)*
Sonar technician (submarine)*

Ship maintenance

Damage controlman
Hull maintenance technician
Instrumentation
Machinery repairman
Molder
Opticalman
Patternmaker

Ship operations

Operations specialist
Quartermaster

Weapons control

Electronics technician
Fire control technician (ballistic missile fire control)*
Fire control technician (gun fire control)*
Fire controlman*

Weapons systems support

Tradevman

8. The Air Force

U.S. Air Force photo

THE AIR FORCE IS UNUSUAL among the armed forces in that its officers are the ones most involved in combat, with enlisted personnel serving a support function on the ground. Without centuries of burdensome tradition and with the officers on the front lines, the Air Force has gained a reputation as the most progressive branch of the service.

Though smaller than the Army or the Navy, the Air Force is in charge of all land-based missiles and strategic bombers, two-thirds of the nation's nuclear arsenal.

The Air Force has over seven thousand aircraft and a total of 604,000 men and women. Of the 109,000 officers, 11 percent are women. Of the 495,000 enlisted personnel, 12 percent are women. Higher percentages of women serve as enlistees and officers in the Air Force than in any other branch of the service.

The Air Force is the most mobile of the armed forces. Flight crew members, especially, are likely to see at least several countries during their careers. The table lists countries hosting one hundred or more U.S. Air Force personnel; at any given time others are moving from one assignment to another.

History

The youngest of the military branches, the Air Force was established in September of 1947. Before that it was a part of the Army.

Prior to the invention of the airplane, the only aerial equipment in the military was the balloon, which was used to gain a high perch from which to observe enemy positions and to direct artillery fire. So it was that the infant Air Force appeared as the Aeronautical Division of the Army Signal Corps in 1907.

During World War I aircraft took a much greater role in battle, and in 1918 the Army Air Service was established. In 1926 it became the Army Air Corps, and this was its name for fifteen years, until it became the Army Air Forces.

During this time aircraft advanced from the fragile planes of World War I to the fighters and bombers of World War II

Location	*Personnel*
UNITED STATES	
Continental United States	436,300
Alaska	11,000
Hawaii	6,700
WESTERN AND SOUTHERN EUROPE	
Belgium	1,800
Germany	41,300
Greece	2,400
Greenland	300
Iceland	1,300
Italy	5,800
Netherlands	2,000
Norway	100
Portugal	1,300
Spain	5,200
Turkey	3,700
United Kingdom	26,400
EAST ASIA AND THE PACIFIC	
Australia	300
Guam	4,000
Japan and Okinawa	16,500
Philippines	9,100
Korea	11,500
MIDDLE EAST	
Saudi Arabia	200
WESTERN HEMISPHERE OUTSIDE THE UNITED STATES	
Canada	100
Panama	2,500

that fought for air superiority over Germany and Japan. The Army Air Forces dropped the first atomic bombs on Hiroshima and Nagasaki, thus ushering in the atomic age.

Finally, in 1947, the Air Force came into its own. A few years later the Korean Conflict saw the first use of jet aircraft by United States fliers; the top ace of that conflict claimed sixteen enemy planes. The Air Force Academy was opened in 1955, and the Air Force began producing its own leaders.

In Vietnam the Air Force flew some of the longest bombing missions in its history, with B-52s taking off from the Philippines, hitting targets in North Vietnam, and then returning. Air Force pilots stationed in South Vietnam worked with the Army in air strikes on enemy forces.

Since Vietnam the Air Force has concentrated on updating the aging B-52 force and building new weaponry based on high technology.

Length of Hitch

Officers. The obligation of Air Force officers breaks down like this, in years:

	Active Duty	*Ready or Standby Reserve*
U.S. Air Force Academy	5	3
Air Force ROTC	4	4
Officer Training School	4	4

In addition to the initial obligation, other factors can add to the years you owe the Air Force.

- ☐ *Flight training.* Training to become a pilot takes fifty-two weeks, and upon winning your wings you must stay in the Air Force for seven more years. If you become a flight officer, such as a navigator, bombardier, or electronics officer, you owe the Air Force six years.
- ☐ *Graduate school.* If the Air Force sends you to graduate school, you have to stay in uniform three years for every year you go to school, to a maximum of four years in the classroom.
- ☐ *Promotion.* When you accept the rank of captain you commit yourself to two more years; the same goes for major. If you accept promotion to lieutenant colonel or colonel you are obligated to serve three more years.

An important note: These add-ons do not have to be served consecutively.

Enlisted. The Air Force offers two enlistment options, a four-year plan and a six-year plan. Every enlistee is guaranteed the training program that is named in the enlistment papers.

Here's how the plans look:

Active Duty	*Ready or Standby Reserve*
4	4
6	2

If you enlist in the reserve, your active duty lasts only long enough for you to take basic training and job training. You then return to civilian life for the rest of the eight years to attend monthly drill and fifteen days of annual training.

Basic Training

The Air Force has the shortest basic training period of all the services: you're in for only six weeks.

When you first arrive, you and forty or fifty other people are put into a group called a flight. The person doing all of the yelling is the military training instructor.

The Air Force prides itself on being different from the older, more tradition-bound services. The basic training includes no midnight marches, and you spend only one day on the firing range. You are not even issued a pack.

Some things are different in Air Force basic training, but it is just as intense as basic training in the other branches of the service. You start the day with physical training that gets more difficult as time passes and you get in shape. Weight control is important to the Air Force. If you are judged to be overweight and a combination of diet and exercise does not sufficiently reduce your waistline, you will have to go through physical training a second time with a new bunch of recruits, so you will have more time to work off the pounds.

In the classroom you study subjects that focus on the Air Force and others that focus on you. Examples of the former include military justice, discharges, Air Force customs and courtesies, and leaves and passes. Courses with the individual in mind deal with personal finance, drug and alcohol abuse, and health.

A typical day. A day in basic training might go like this:

5:00–6:15 A.M.	Reveille, physical conditioning, and showers
6:15–7:30	Breakfast
7:30–8:30	Dorm preparation for inspection
8:30–11:30	Academic classes
11:30–12:30 P.M.	Lunch
12:30–4:30	Academic classes, drill practice, personal inspection, and retreat
4:30–5:30	Dinner
5:30–6:30	Mail call, briefings
6:30–9:00	Dorm preparation, study time, personal hygiene
9:00	Lights out

All enlisted personnel take basic training at Lackland Air Force Base near San Antonio, Texas. Men and women undergo identical training. They live in different wings of the same dormitory—which is air-conditioned—and are assigned to different flights in the same squadron.

Jobs

Choosing a job begins with the ASVAB, the test you take before joining the Air Force. Once you have an idea of where your talents lie, you can sit down with a classifier and work out a plan.

Women have particularly good opportunities in the Air Force. Since few enlisted personnel are in combat, the percentage of job categories barred to women, according to the Department of Defense, is only about 2 percent—by far the lowest in all of the service.

You can choose a particular job, or, if you aren't ready to make up your mind, you can sign up for a general career area. The areas are:

☐ *Administrative Aptitude Area.* This includes positions such as radio operator, freight traffic specialist, and accounting specialist.
☐ *Electronics Aptitude Area.* In this group you will find missile electronics, radar and computer systems, and instrumentation systems.
☐ *General Aptitude Area.* This area covers fireman, printer, computer operator, and medical specialist.
☐ *Mechanical Aptitude Area.* Here you might find yourself doing aircraft maintenance, welding, carpentry, or even piloting a bulldozer.

You may have a hard time making up your mind, but it's generally better to do so *before* you enlist. That way you can make sure you get the job training you really want. Once you're in you are bargaining from a weaker position.

For those who decide to go for a specific job, the choices are wide-ranging. The Air Force offers approximately two hundred forty occupations. Here are two of them, along with descriptions, length of training required, and civilian equivalents.

☐ *Cable and antenna installation/maintenance specialist.* People in this job install, maintain, and repair pole lines and aerial cables. Climbing towers and poles is required. They work with underground cable and field wire and related electrical equipment such as control cables, support structures, radomes, and transmission lines.

Following basic training the schooling takes approximately thirteen weeks at Sheppard Air Force Base near Wichita Falls, Texas. The civilian equivalent is a lineman for a telephone company, a cable television installer, or a lineman for a radio and television company.
☐ *Management analysis specialist.* This job consists in compiling and preparing summaries of statistical data reflecting actual versus planned performance toward Air Force objectives. A person in this job compares accomplishments with objectives, determines trends, and maintains data banks.

The training takes nine weeks at Sheppard Air Force Base in Texas.

Bonuses

Like the other branches of the service, the Air Force uses cash bonuses to entice recruits to fill certain jobs. Many of the positions require extensive training—some more than a year—and to get the bonus you sign up for a longer hitch.

The specialties that offer bonuses change from time to time as the needs of the Air Force change. To find out what is being offered now you should check with an Air Force recruiter. To give you an idea of what these jobs might be, here are some specialties that rated bonuses in the past, as well as the amounts of those bonuses.

☐ *Cryptologic linguist specialist, $2,000.* This job consists in listening to foreign radio transmissions and recording signals of interest to the Air Force. The amount of training required varies, but it could take as long as seventy-eight weeks. Most personnel with this job are sent overseas.

☐ *Explosive ordnance disposal specialist, $1,000.* This job consists in taking unexploded or damaged missiles or bombs and making them safe. The weapons may be explosive, chemical, biological, or nuclear. If any of the material inside the weapons escapes you must know how to clean it up. The training takes nineteen weeks.

Job List

Following is a list of most Air Force occupations. Like civilian occupations, each has several levels of expertise. A person in the Air Force usually starts as a helper, moves into an apprenticeship, becomes a specialist, then a supervisor, and finally a manager. The job titles below are entry-level. Those jobs barred to women are marked with asterisks.

Accounting and finance, auditing

Financial management helper
Financial services helper
Auditing helper

Administration

Chapel management helper
Administration helper
Reprographic helper

Legal services helper

Aircraft maintenance

Helicopter helper
Tactical aircraft maintenance helper
Strategic aircraft maintenance helper
Airlift aircraft maintenance helper
General aircraft maintenance helper
0-2 aircraft airframe and power plant helper

Aircraft systems maintenance

Aircraft electrical systems helper
Aircraft environmental systems helper
Aircrew egress systems helper
Aircraft fuel systems helper
Aircraft pneudraulic systems helper
Aerospace ground equipment helper
Jet engine helper
Turboprop propulsion helper
F-100 jet engine helper
Machine shop helper
Corrosion control helper
Nondestructive inspection helper
Fabrication and parachute helper
Metals processing helper
Airframe repair helper

Aircrew operations

Defensive aerial gunner helper
Inflight refueling operator helper
Flight engineer helper (helicopter)
Flight engineer helper (performance qualified)
Aircraft loadmaster helper
Pararescue/recovery helper
Airborne communications systems helper
Airborne warning coding and communications systems helper
Airborne computer systems helper
Airborne coding and communications equipment helper

Aircrew protection

Survival training helper
Aircrew life support helper

Audiovisual

Audiovisual media helper
Graphics helper
Still photographic helper
Audiovisual production-documentation helper
Imagery production helper

Avionic systems

Bomb-navigation systems helper
Defense fire control systems helper
Weapon control systems helper*
Avionic sensor systems helper
Offensive avionics systems helper
Aircraft computer and multiplexing systems helper
Defensive avionics systems helper
Precision measurement equipment laboratory helper
Automatic flight control systems helper
Avionics instruments systems helper
International avionics EW equipment and computer helper
International avionics computerized test stations and computer helper
International avionics manual test station and computer helper
International avionics attack control systems helper
International avionics instruments and flight control systems helper
International avionics communication, navigation, and pen-aids systems helper
Avionic communications helper

Avionic navigation systems helper
Airborne warning and control radar helper
EW systems helper
Avionic inertial and radar navigational systems helper
Airborne command post communications equipment helper

Band

Clarinet player
Saxophone player
Bassoon player
Oboe player
Flute or piccolo player
French horn player
Cornet or trumpet player
Baritone or euphonium player
Trombone player
Tuba player
Percussion player
Piano player
Guitar player
Music arranger
Vocalist
Electric bass/string bass player
Air National Guard band
Instrumentalist helper

Command control systems operations

Airfield management helper
Operations resources management helper
Air traffic control helper
Combat control helper
Command and control helper
Tactical air command and control helper
Aerospace control and warning system helper
Space systems operations helper
Weather equipment helper

Communications-electronics systems

Airborne MET/ARE helper
Air traffic control radar helper
Automatic tracking radar helper
Wideband communications equipment helper
Navigation aids equipment helper
Ground radio communications helper
Television equipment helper
Space communications systems equipment operator/helper
Electronic computer and switching systems helper
Electronic computer and cryptologic equipment systems helper
Telecommunications systems maintenance helper
Space systems equipment maintenance helper

Contracting

Contracting helper

Cost and management analysis

Cost and management analysis helper

Dental

Dental assistant helper
Dental laboratory helper

Education and training

Education helper
Training helper
Instructional systems helper
Combat arms training and maintenance helper
Gunsmith helper

Fire protection

Fire protection helper

Food services

Food service helper

Fuels

Fuel helper

Geodetic

Geodetic helper

Information systems

Information systems operation helper
Programming helper
Information systems radio operator helper
Information systems electromagnetic spectrum management helper
Information systems control helper
Information systems programming management helper

Intelligence

Intelligence operations helper
Target intelligence helper
Radio communications analysis helper
Linguist/interrogator helper
Electronic intelligence operations helper
Imagery interpreter helper
Morse systems helper
Printer systems helper
Germanic cryptologic linguist helper (German, Dutch, Flemish, or Swedish)
Romance cryptologic linguist helper (Spanish, Portuguese, French, Italian, or Romanian)
Slavic cryptologic linguist helper (Russian, Polish, Czech, Serbo-Croatian, White Russian, Hungarian, Lithuanian, Slovenian, or Bulgarian)
Far East cryptologic linguist helper (Mandarin Chinese, Vietnamese, Thai, Cambodian, Lao, Japanese, Korean, or Cantonese Chinese)
Mid-East cryptologic linguist helper (Arabic, Syrian Arabic, Hebrew, Persian, Turkish, Greek, Indonesian, or Hindi-Urdu)

Instrumentation

Instrumentation helper

Intricate equipment maintenance

Precision imagery and audiovisual media maintenance helper
Aerospace photographic systems helper

Logistics plans

Logistics plans helper

Maintenance management systems

Maintenance data systems analysis helper
Maintenance scheduling helper

Marine

Seaman helper
Marine engine helper

Mechanical electrical

Electrical helper
Electric power line helper
Electric power production helper
Refrigeration and cryogenics helper
Liquid fuel systems maintenance helper
Heating systems helper
Civil engineer control systems helper

Medical

Medical service helper*
Cardiopulmonary laboratory helper
Surgical service helper*
Radiologic helper
Nuclear medicine helper
Pharmacy helper
Medical administration helper
Bioenvironmental engineering helper
Environmental medicine helper
Aerospace physiology helper
Optometry helper (ophthalmology)*
Physical therapy helper
Occupational therapy helper
Mental health clinic helper
Mental health unit helper
Medical material helper
Biomedical equipment maintenance helper
Orthotic helper
Medical laboratory helper
Histopathology helper
Cytotechnology helper
Diet therapy helper

Missile systems maintenance

Missile systems maintenance*
Laser/electric operation weapon kits
Missile maintenance helper*
Missile facilities helper*
Missile pneudraulic helper
Missile liquid propellant systems maintenance helper

Morale, welfare, and recreation

Fitness and recreation helper
Open mess management helper

Munitions and weapons maintenance

Munitions systems helper
Aircraft armament systems helper
Nuclear weapons helper
Explosive ordnance disposal helper

Personnel

Personnel helper
Personal affairs helper
Career advisory helper
Manpower management helper
Social actions helper (equal opportunity/human relations)
Social actions helper (drug/alcohol abuse control)

Public affairs

Public affairs helper
Radio and TV broadcasting helper
Historian helper

Safety

Safety helper
Disaster preparedness helper

Sanitation

Pest management helper
Environmental support helper

Security police

Security helper
Law enforcement helper

Services

Services helper
Meatcutting helper
Subsistance operations helper

Special investigations

Special investigations helper

Structural/pavements

Pavements maintenance helper
Construction equipment helper
Carpentry helper
Masonry helper
Metal fabricating helper
Protective coating helper
Plumbing helper
Engineering assistant's helper
Civil engineering management helper
Production control helper

Supply

Inventory management helper
Materiel facilities helper
Supply systems analysis helper

Training devices

Defensive systems trainer helper
Flight simulator helper
Navigational/tactical training divisions helper
Missile trainer helper

Transportation

Passenger and household goods
Freight traffic helper
Packaging helper
Vehicle operator/dispatcher helper
Air passenger helper
Air cargo helper

Vehicle maintenance

Special purpose vehicle and equipment mechanic's helper
Special vehicle mechanic's helper (firetrucks)
Special vehicle mechanic's helper (refueling vehicles)
General purpose vehicle maintenance helper
Vehicle body maintenance helper
Vehicle maintenance construction and analysis helper

Weather

Weather helper

Wire communications systems maintenance

Cable and antenna systems installation/maintenance helper
Cable splicing installation and maintenance helper
Telephone central office switching equipment helper, electrical/electromechanical
Missile control communications systems helper
Telephone equipment installation and repair helper

9. The Marine Corps

U.S. Marine Corps photo

If one word could be used to describe the Marine Corps, that word would be *versatile.* Marines have the longest basic training of all the services, and they are prepared to fight under a variety of conditions by themselves or in close cooperation with the Army, Navy, or Air Force.

The Corps prides itself on being able to deploy troops on short notice, and for this reason Marines are known as the "first to fight."

The smallest of the Department of Defense branches, the Marine Corps has 196,000 men and women in uniform. Three percent of the 20,000 officers are women, as are 5 percent of the 177,000 enlisted forces. These are the lowest percentages of women in any branch of the service.

Of all the branches, the Marine Corps offers the greatest number of places to which you can travel. There are Marines at every U.S. embassy and at over one hundred fifty naval

Location	*Personnel*
UNITED STATES	
Continental United States	149,000
Alaska	200
Hawaii	8,900
Afloat	900
WESTERN AND SOUTHERN EUROPE	
Germany	100
Iceland	100
Italy	300
Spain	200
United Kingdom	400
Afloat	2,500
EAST ASIA AND THE PACIFIC	
Guam	400
Japan and Okinawa	21,500
Philippines	1,200
Korea	1,000
Afloat	1,800
WESTERN HEMISPHERE OUTSIDE THE UNITED STATES	
Cuba (Guantanamo Bay)	600
Panama	200
Puerto Rico	300

installations around the world. The table omits the embassies, since only countries in which there are more than one hundred Marines are included. At any given time, too, a certain number of Marines are between assignments.

History

The Marines have always been linked to the Navy. The Continental Congress established the Marine Corps in 1775, and for a time Marines served only on ships. When an American warship would engage in battle with that of an enemy, the Marines would swarm aboard the other vessel and engage in hand-to-hand combat.

It was not until 1834 that separate companies were formed on land, and thereafter the Marines could fight alongside the Navy and the Army. The Corps has participated in every U.S. war, including the War of 1812, the Mexican War (from which the line about the "halls of Montezuma" in the Marines' Hymn comes), and the Spanish-American War.

The Marines fought well in World War I, but it was with the many amphibious landings of World War II that the Corps truly distinguished itself. Having the capability to come in from the sea, establish a beachhead, and keep moving, the Marines led the American advance toward Japan.

Since World War II the role of the Marines has expanded. They fought alongside the Army in Korea and Vietnam, and they have been sent by presidential order to such places as Grenada and Beirut, Lebanon.

Currently the Marines guard U.S. embassies, naval installations, and ships, but their primary function is readiness for war. "Let it be known," said the commandant of the corps, "that, when called, Marine forces will fight our country's battles in the configuration most useful to our nation."

Length of Hitch

Officers. All commissioned officers are obliged to serve the Corps for a specific number of years. After that they can resign and return to civilian life or serve in the Ready Reserve.

The obligations break down like this, in years:

	Active Duty
U.S. Naval Academy	5
Naval ROTC	4
Officer Candidate Class	3

The three- to five-year obligation is for your initial officer training. If you want more training from the Corps, the opportunity is there—but your obligation will increase.

Here are some of the opportunities for officers along with the increased obligations.

☐ *Graduate school.* If you want to get an advanced degree and the Marine Corps sends you to graduate school, you have to stay in uniform three years for the first twelve months of schooling and one more year for any time over twelve months, to a maximum of eighteen months. If, for example, you enter a fifteen-month master's program in business administration, you will owe the Marines four years.

☐ *Promotion.* If you accept a promotion to the next rank you commit yourself to two more years.

☐ *Aviation training.* In the Marines you can fly jet fighters, helicopters, or multi-engine transports. Pilot training (under Navy supervision) takes about one and a half years, and once you get your wings you owe the Marine Corps four and a half years.

An important note: These add-ons are not always cumulative. This means if you were to go through Naval ROTC and then through pilot training you would not owe the Marine Corps ten and a half years. If you entered flight training as soon as possible, your obligation after winning your wings would be only four and a half years.

Enlisted. Men who join the enlisted ranks of the Marine Corps sign up for eight years. Women, unlike men, do not have to serve in the reserve after leaving active duty. How a man's eight years are spent, however, depends on the program he selects.

Here are the options, in years, for the active Marines:

Active Duty	*Ready or Standby Reserve*
3	5
4	4
6	2

And for the Marine Corps Reserve—

	Active Duty	*Ready Reserve*	*Standby Reserve*
Women only	⅓–1	3	5
Men and women	⅓–1	4	3
Men and women	⅓–1	8	0

The length of active duty required of those enlisting in the reserves depends on the kind of schooling taken after basic training.

Basic Training

While the other branches of the military soft-pedal their basic training, the Marines thrive on their tough reputation. Basic training, or boot camp, lasts for ten weeks—the longest in the service—and is taken either at Parris Island, South Carolina, or in San Diego, California.

Women's basic training is offered only at Parris Island and lasts for eight weeks. Women have a shorter training period because they are barred from taking part in combat and therefore do not have to learn battlefield skills.

A typical day. This might be your schedule one day during the seventh week of boot camp.

5:00 A.M.	Reveille
5:00–6:30	Breakfast and cleanup
6:30–9:00	Conditioning march

9:00–9:20	Recovery from physical training
9:30–11:30	Instruction in the protective mask
11:30–12:00 P.M.	Lunch
12:00–2:00	Instruction in nuclear, biological, and chemical defense
2:00–3:30	Gas chamber exercise. Simulated nuclear, biological, and chemical warfare exercise
3:30–4:30	Instruction in camouflage, cover, and concealment
4:30–5:30	Instruction in estimating combat situations in various terrains
5:30–6:00	Dinner
6:00–7:30	(Decided by drill instructor)
7:30–7:50	Hygiene inspection
7:50–8:50	Free time
8:50–9:00	Final muster
9:00	Taps

The physical part: men. The exercises, running, and seemingly endless pushups culminate in the Physical Fitness Test, in which men must complete minimum requirements in three events: the pullup, the bent-knee situp, and the three-mile run. Recruits aren't the only ones who have to take this test. All Marines—officers and enlisted—take it every six months; the Corps believes in keeping fit.

Here are the minimum acceptable performances:

Age	*Pullups*	*Situps*	*3-Mi. Run (min.)*	*Passing Score*
17–26	3	40	28	135
27–39	3	35	29	110
40–45	3	35	30	85

You get extra points for any pullups or situps above the minimum and for a shorter time on the three-mile run. You have to complete the minimum number in each group and then get enough extra points to meet the passing score. One hundred thirty-five is barely passing; to get a first-class rating you have to score 225.

The physical part: women. The system is the same as the men's, although the events are different. Instead of pullups women perform a flexed-arm hang, and instead of a three-mile run women have to cover only one and a half miles.

Here is the point system:

Age	*Flexed-Arm Hang (sec.)*	*Situps*	*1½-Mi. Run (min.)*	*Passing Score*
17–25	16	22	15	100
25–31	14	20	16	82
32–38	12	18	17	64

One hundred points is barely passing; to score first class you must have 200 points.

Basic training for men and women ends when they successfully qualify in three of the following: an inspection by the battalion commander, the physical fitness test, a comprehensive final exam on Marine matters, and marksmanship.

After graduation the trainee is no longer a recruit; he or she will now be addressed as a United States Marine.

Jobs

The Marine Corps has thirty-eight occupational fields comprising over three hundred jobs that are open to enlistees right after boot camp, including a large number of combat positions. Like the Army, the Marines call each job a Military Occupational Specialty, or MOS.

You and the classifier sit down with your ASVAB scores and figure out the jobs for which you qualify. Once you arrive at a decision, the Marine Corps guarantees you'll get the training. After boot camp you'll report to one of the Marine schools across the country and begin training.

Here are three career fields and some of the jobs that fall in them as well as the length of training for each.

☐ *Audiovisual.* Marines in this job field operate still and motion picture equipment, process film, and print photographs. They may also repair cameras and edit films.

Graphics specialist	12 weeks
Audio/TV production specialist	12 weeks
Still photo specialist	12 weeks

☐ *Utilities.* Men and women in this field install, operate, and maintain electrical, water supply, heating, plumbing, and refrigeration and air conditioning equipment.

Basic refrigeration mechanic	7 weeks
Basic electrician	7 weeks
Basic plumbing and water supplyman	7 weeks

☐ *Public Affairs.* These Marines gather information and write news stories, historical pieces, and radio and television scripts.

Information specialist (broadcaster)	10 weeks
Information specialist (journalist)	10 weeks

Advanced Training

The Marine Corps has an apprenticeship program similar to that of the Army. If you are on active duty and your MOS is approved for the program, you can log the hours you spend on the job and emerge from the Corps with journeyman status. This could pave the way for high-paying jobs in the civilian world.

In addition, the Marine Corps offers advanced training in most of its MOS's. As Marines rise in rank and move into more supervisory positions, both officers and enlisted personnel are sent to classes in management. They also find themselves serving as class instructors, thus learning more about their field.

Bonuses

Like the other branches of the service, the Marine Corps gives bonuses to enlistees who select jobs in which the Corps has a lot of vacancies. These are the bonuses that have been offered in the past year: $7,000 for enlistment as artillery weapon/

turret repairer (male only), ammunition technician, electronics maintenance, or air traffic controller–tower; $4,500 for enlistment as air delivery specialist (male only), field artillery fire control (male only), military police, or correctional specialist (male only). For an up-to-date list of jobs paying bonuses, speak with a Marine Corps recruiter.

Job List

Here is a list of most of the jobs in the Marine Corps, in categories defined by the Corps. Jobs barred to women are marked with asterisks.

Air control/air support/ anti-air warfare

Basic air control/air support/anti-air warfare Marine
FAAD gunner
HAWK missile system operator
Air command and control electronics operator
Tactical air defense controller
Air support operations operator

Air traffic control and enlisted flight crews

Basic air traffic controller/enlisted flight crew Marine
Air traffic controller—trainee
Air traffic controller—tower
Air traffic controller—radar
Air traffic controller*
Aerial navigator—trainee
First navigator
Airborne radio operator/ loadmaster—trainee
Airborne radio operator/loadmaster

Airfield services

Basic airfield services Marine
Aircraft recovery specialist
Aviation operations specialist
Aircraft firefighting and rescue specialist

Aircraft maintenance

Basic aircraft maintenance Marine
Aircraft mechanic—trainee
Aircraft mechanic
Aircraft power plants mechanic
Aircraft flight engineer, KC-130—trainee
Aircraft flight engineer, KC-130
Spectrometric oil analysis technician
Aircraft power plants test cell operator, fixed wing
Maintenance specialist
Aircraft welder
Aircraft non-destructive inspection technician
Aircraft maintenance administration clerk
Aircraft maintenance data analysis technician
Aircraft maintenance computer systems analyst/operator
Aircraft hydraulic/pneumatic mechanic—trainee
Aircraft hydraulic/pneumatic mechanic
Aircraft airframes maintenance chief
Flight equipment Marine
Aircraft maintenance ground support equipment mechanic (GSE)—trainee

Aircraft maintenance GSE mechanic
Cryogenics equipment operator
Aircraft maintenance GSE hydraulic/pneumatic/structures mechanic
Aircraft maintenance GSE electrician
Aircraft maintenance GSE refrigeration mechanic
Aircraft maintenance GSE chief
Aircraft safety equipment mechanic—trainee
Aircraft safety equipment mechanic
Aircraft structures mechanic—trainee
Aircraft structures mechanic
Helicopter mechanic—trainee
Helicopter mechanic
Helicopter power plants mechanic
Helicopter dynamic components mechanic
Aircraft power plants test cell operator, rotary wing
Helicopter structures mechanic
Helicopter hydraulic/pneumatic mechanic
Helicopter airframes maintenance chief
Presidential support specialist
Helicopter crew chief

Ammunition and explosive ordnance disposal

Basic ammunition and explosive ordnance disposal Marine
Ammunition technician
Explosive ordnance disposal technician
Ground nuclear ordnance technician

Auditing, finance, and accounting

Basic auditing, finance, and accounting Marine
Personal financial records clerk
Travel clerk
Auditing technician
Accounting technician

Aviation ordnance

Basic aviation ordnance Marine
Aviation ordnance trainee
Aviation ordnance munitions technician, IMA
Aircraft ordnance technician
Aviation ordnance equipment repair technician, fixed wing
Aviation ordnance equipment repair technician, rotary wing, IMA
Marine wing weapons unit specialist

Avionics

Basic avionics Marine
Aircraft communications/navigation systems technician—trainee
Aircraft communications/navigation systems technician
Aircraft electrical systems technician—trainee
Aircraft electrical systems technician
Advanced avionics technician—trainee
Aircraft weapons systems specialist
Aircraft radar reconnaissance systems technician
Aircraft communications/navigation/radar systems technician
Aircraft integrated weapon systems technician
Aerial camera systems technician—trainee
Aerial camera systems technician
Imagery interpretation equipment repair technician
Aircraft electronic countermeasures systems technician

Aircraft communications systems technician, fighter/attack
Aircraft communications systems technician, helicopter
Aircraft cryptographic systems technician
Aviation electronic micro-miniature repair technician
Aircraft electrical/instrument systems technician
Aircraft flight control and air data computer systems technician
Aircraft analog display system SACE technician
Aircraft ballistics computer technician
Aircraft radar systems technician
Aircraft inertial navigation systems technician
Semi-automatic checkout equipment
Aircraft ALQ-99 jammer systems technician
Aircraft ALQ-99 track and surveillance systems technician
Aircraft ALQ-99 display/AYA-6 systems technician
ECM module repair technician
Avionics test set (ATS) technician
Radar test station (RTS) technician
Carrier aircraft inertial navigation system (CAINS) technician
Hybrid test set technician
Aircraft weapons systems technician
Aircraft radar/IR reconnaissance systems technician
Aerial camera, ADAS systems technician
Aircraft forward looking infrared/TOW technician
Aircraft deceptive electronic countermeasures systems technician
Aircraft passive electronic countermeasures systems technician
Aircraft active electronic countermeasures systems technician
Aviation PME calibration technician
Aviation PME and ATE repair technician

Band

Basic musician
Musician, oboe/English horn
Musician, bassoon
Musician, clarinet
Musician, flute and piccolo
Musician, saxophone
Musician, cornet/trumpet
Musician, baritone horn/euphonium
Musician, French horn
Musician, trombone
Musician, tuba and string bass/electric bass
Musician, percussion (drums, timpani, and mallets)
Musician, piano or accordion or guitar
Arranger, band
Drum and bugle corps drum major
Bugler, soprano or mellophone
Bugler, French horn
Bugler, bass baritone
Bugler, contrabass
Drum and bugle corps arranger
Drummer, drum and bugle corps

Category "B" MOS's

Quality assurance technician (subsistence)
Guard
Education assistant
Recruiter
Psychological operations NCO
Civil affairs noncommissioned officer
Drill instructor
Marksmanship instructor

Scout sniper
Water safety/survival instructor
Interpreter (designated language)
Surveillance sensor operator
Surveillance sensor maintenance man
Reconnaissance man, parachute jump qualified
Reconnaissance man, SCUBA qualified
Reconnaissance man, parachute and SCUBA qualified
Infantry operations specialist
Firefighter
Barracks and grounds Marine
Food service attendant
Athletic and recreation assistant
Military affiliate radio system radio operator
Graves registration specialist

Data/communications maintenance

Basic data/communications maintenance Marine
Telephone technician
Cable systems technician
Central office installer-repairer
Teletype technician
KG-13 teletype technician
Electronic switching equipment technician
Technical controller
Fixed ciphony technician
Mobile data terminal technician
KW-26 terminal technician
Mobile communication central technician
Microwave equipment repairer
Fleet satellite terminal technician
Ground mobile forces SatCom technician
Ground radio repairer
Radio technician
Test measurement and diagnostic equipment technician
Metrology technician
Small missile systems technician
Radiac instrument technician
Communication security equipment technician
KG-30 COMSEC technician
Ground radar repairer
FADAC radar repairer
Artillery electronics repairer
Weapons location equipment repairer
Ground radar technician
Data/communications maintenance chief

Data systems

Basic data systems Marine
Network control specialist
Computer operator
Data control coordinator
Programmer, COBOL
Programmer, ALC
Programmer, EDL

Drafting, surveying, and mapping

Basic drafting, surveying, and mapping Marine
Construction drafter
Map compiler
Construction surveyor
Geodetic surveyor

Electronics maintenance

Basic electronics maintenance Marine
Microminiature circuit repair specialist
Improved HAWK fire control repairer
Improved HAWK information coordination central repairer
Improved HAWK K firing section repairer
Improved HAWK pulse radar technician
Improved HAWK continuous wave radar technician

Improved HAWK automatic fire distribution and engagement simulator system technician
Improved HAWK fire control technician
Improved HAWK missile system maintenance technician
Improved HAWK mechanical system repairer
Aviation radio repairer
Aviation meteorological equipment technician
Aviation radio technician
Aviation radar repairer
Aviation fire control repairer
Aviation radar repairer
Aviation fire control technician
Aviation radar technician
Air traffic control navigational aids technician
Air traffic control radar technician
Air traffic control communications technician
Air traffic control systems maintenance chief
Tactical air command central repairer
Tactical air operations central repairer
Tactical data communications central repairer
Digital data systems technician, Honeywell

Engineer, construction and equipment

Basic engineer, construction, and equipment Marine
Metal worker
Engineer equipment mechanic
Engineer equipment operator
Rock quarry operator
Combat engineer
Bulk fuel specialist

Field artillery

Basic field artillery man*
Field artillery cannoneer*
Field artillery nuclear projectileman*
Field artillery radar operator*
Field artillery fire control man*
Artillery meteorological man*
Fire support man*

Food service

Basic food service marine
Baker
Cook, specialist
Food service specialist

Identifying MOS's and reporting MOS's

Basic Marine, general service
Special assignment
Billet designator
College degree
Special technical operations Marine (officer, enlisted)
SCUBA Marine
Parachutist/SCUBA Marine
Ground safety specialist
Parachutist
Basic Marine with enlistment guarantee
Tactical data systems specialist

Infantry

Basic infantryman*
Rifleman*
LAV assaultman*
LAV crewman*
Reconnaissance man*
Machine gunner*
Mortar man*
Assaultman*
Antitank/assault guided missile man*

Intelligence

Basic intelligence Marine
Intelligence specialist
Interrogation-translation specialist

Legal services

Basic legal services Marine
Legal services specialist
Legal services notereader/transcribe (stenotype)

Logistics

Basic logistics Marine
Maintenance management specialist
Logistics/embarkation specialist
Air delivery specialist
Landing support specialist

Marine Corps exchange

Basic Marine Corps exchange Marine
Exchange Marine

Military police and corrections

Basic military police and corrections Marine
Military police
Military police dog handler
Correctional specialist

Motor transport

Basic motor transport Marine
Body repair mechanic
Organizational automotive mechanic
Intermediate automotive mechanic
Vehicle recovery mechanic
Fuel and electrical systems mechanic
Crash/fire/rescue vehicle mechanic
Motor vehicle operator
Tractor trailer operator
Sèmi-trailer refueler operator

Nuclear, biological, and chemical

Basic nuclear, biological, and chemical Marine
Nuclear, biological, and chemical defense specialist

Operational communications

Basic operational communicator
Field wireman
Construction wireman
Field radio operator
Microwave equipment operator
Radio telegraph operator
High frequency communication central operator
Fleet SATCOM terminal operator
Ground mobile forces SATCOM operator
Communication center operator

Ordnance

Basic ordnance marine
Infantry weapons repairer
Rifle team equipment repairer
Artillery weapons/turret repairer
Tracked vehicle repairer, assault amphibian vehicle
Tracked vehicle repairer, self-propelled artillery
Tracked vehicle repairer, tank
Tank turret repairer
Light armored vehicle repairer
Repair shop machinist
Optical instrument repairer
Electro-optical equipment technician

Personnel and administration

Basic administrative Marine
Personnel clerk
Unit diary clerk
Administrative clerk
Postal clerk
Administrative control unit specialist

Printing and reproduction

Basic printing and reproduction Marine
Offset press operator
Process camera operator
Reproduction equipment repairer

Public affairs

Basic public affairs Marine
Broadcast journalist
Print journalist
Photojournalist

Signals intelligence/ ground electronic warfare

Basic signals intelligence/ground electronic warfare operator
Manual Morse intercept operator
Signals intelligence analyst
Non-Morse intercept operator/analyst
Special intelligence communicator
Cryptologic support specialist
Cryptologic linguist, Middle Eastern
Cryptologic linguist, Chinese
Cryptologic linguist, Korean
Cryptologic linguist, Spanish
Cryptologic linguist, Russian

Supply administration and operations

Basic supply administration and operations Marine
Supply administration and operations clerk
Warehouse clerk
Packaging specialist
Subsistence supply clerk
Aviation supply clerk
Automated information system (AIS) computer operator

Tank and assault amphibian

Basic tank and assault amphibian crewman*
Tank crewman*
Assault amphibian crewman*

Training and audiovisual support

Basic training and audiovisual support Marine
Graphics specialist
Training and audiovisual operations specialist
Photographer
Photographic technician
Audiovisual equipment technician
Combat photographer/motion media

Transportation

Basic transportation Marine
Traffic management specialist

Utilities

Basic utilities Marine
Electrician
Electrical equipment repair specialist
Refrigeration mechanic
Hygiene equipment operator
Well driller
Fabric repair specialist
Office machine repair specialist

Weather service

Basic weather service Marine
Weather observer
Rawinsonde operator

10. The Coast Guard

U.S. Coast Guard photo

THE COAST GUARD IS UNIQUE among the armed forces in three ways. First, it does not report to the Pentagon. It is administered instead by the Department of Transportation—except in times of war, when it operates in conjunction with and under the command of the Navy.

The second difference concerns the mission of the Coast Guard. The other branches of the service are constantly training and preparing themselves for combat, something that, fortunately, doesn't occur very often. The typical Air Force pilot and Army Ranger will probably never attack hostile forces. Men and women in the Coast Guard, however, train for duties they will perform all the time.

The third unique characteristic of the Coast Guard is this: alone among the armed forces, it bars none of its jobs to women. Both male and female members conduct search and rescue missions, enforce customs and fishing laws, combat drug smuggling, maintain lighthouses and other navigational aids, control pollution, break ice, and promote boating safety. These operations are performed on lakes and rivers as well as the high seas. A person in the Coast Guard might instruct owners of small boats about safety on the Great Lakes or serve on an icebreaker in the arctic. The motto of the Coast Guard is *Semper Paratus*—"Always Ready."

The Coast Guard has 38,000 men and women in uniform, 32,000 enlistees and 6,000 officers. Women make up almost 8 percent of the enlisted ranks and almost 3 percent of the officer corps. Although the bulk of the Coast Guard's operations are conducted in and around the United States, personnel are stationed in Japan, various islands in the Pacific, Australia, Iceland, Norway, England, France, Spain, Italy, Turkey, Jamaica, Barbados, Colombia, the Virgin Islands, Liberia, Haiti, and Argentina.

History

On August 4, 1790, Congress authorized the construction of ten boats to combat smuggling. The young republic needed all the tax money it could get, and the Revenue Marine, as the fleet was called, was virtually the only oceangoing force the country had for eight years.

With the exception of the short war on the Tripoli pirates, the Revenue Marine participated in every American war. The cutter *Harriet Lane* fired the first naval shot of the Civil War off Fort Sumter, South Carolina, and the Revenue Service fought in the Spanish-American War in Cuba and the Philippines.

In 1915 the Revenue Cutter Service, as it had come to be known, merged with the U.S. Lifesaving Service to form the Coast Guard. World War I soon followed, and the Coast Guard was called upon to escort ships across the Atlantic and to patrol the U.S. coasts. In proportion to its size the Coast Guard suffered the highest loss of life of any branch of the service in World War I.

Between the wars the Coast Guard returned to its peacetime duties of saving lives, enforcing laws, and aiding navigation. During Prohibition in the 1920s it was charged with preventing people from smuggling alcoholic beverages. This was a thankless task, but it served to build up the Coast Guard to three times its World War I size.

This proved enormously helpful during World War II, when the Coast Guard once again operated as a part of the Navy. Antisubmarine warfare was a prime concern, with U-boats prowling the country's eastern coast, and the Coast Guard once more helped escort convoys across the North Atlantic. The small-boat experience of the Coast Guard was put to use in the island hopping of the Pacific campaign.

The Coast Guard participated in the Korean Conflict, and in Vietnam it worked to prevent the North Vietnamese from bringing in men and supplies by sea.

Since then the Coast Guard has assumed the tasks of combating drug smuggling, enforcing the boundaries of increased ocean territory of the United States, and investigating water pollution. All this is in addition to the full-time work of saving lives, maintaining aids to navigation, breaking ice, and promoting boating safety.

Length of Hitch

Officers. All officers in the Coast Guard have to serve a minimum of eight years in a combination of active and reserve duty. The obligation breaks down like this:

	Active Duty	*Reserves*
Coast Guard Academy	5	3
Officer candidate school	3	5

Certain things add to an officer's obligation in the Coast Guard:

☐ *Graduate school.* If the Coast Guard sends you to graduate school you must stay in uniform two years for every year spent in school.
☐ *Promotion.* Although promotions occurring during your initial period of obligation do not increase that obligation, for an elevation in rank thereafter you owe the Coast Guard two years.
☐ *Aviation training.* The obligation for flight training is the longest in the Coast Guard. The training takes approximately one and a half years, and at the completion of training you owe the Coast Guard five more years.

An important note: These add-ons are not necessarily consecutive. If, for example, you finished flight training and received a promotion at the same time, you would not owe the Coast Guard seven years—only the five for flight training.

Enlisted. Everyone who enlists in the Coast Guard must spend eight years in a combination of active and reserve duty. It breaks down like this:

	Active Duty	*Ready Reserve*	*Standby Reserve*
ACTIVE COAST GUARD	4 years	4 years	0 years
COAST GUARD RESERVES			
Reserve Split Training Program (boot camp one summer, job training the next)	9 weeks of basic training; seven weeks of job training	7½	0
Petty Officer Selectee Program	24 weeks of boot camp and job training	7½	0
On-the-Job Training Program	8 weeks of boot camp; 4 weeks of on-the-job training	7¾	0
Direct Petty Officer Program (for men and women 26–35 years old with a skill the Coast Guard wants)	2 weeks of orientation	8	0

In any program you can transfer from the Ready Reserve to the Standby Reserve with the approval of your commanding officer. You serve no duty in the Standby Reserve, and you get no pay.

Basic Training

Groups of approximately sixty men and women make up a company, which trains under the supervision of the company commander. Basic training, or boot camp, takes place at Cape May, New Jersey, and lasts for nine weeks.

Recruits quickly learn that although the Coast Guard is not a part of the Department of Defense, the basic training is conducted in a thoroughly military manner.

A typical day. Coast Guard basic training incorporates courses common to all the services and adds a special emphasis on seamanship and activities that relate to the Coast Guard's peacetime mission.

A day in the third week of boot camp shapes up like this:

5:15 A.M.	Reveille
5:30–6:00	Physical training
6:00–6:30	Breakfast
6:30–7:30	Barracks cleanup and inspection
7:30–8:20	(Company commander decides)
8:30–9:20	Human relations course
9:30–11:20	Swimming
11:30–12:20 P.M.	Lunch
12:30–2:20	Military conduct class
2:30–4:20	Communications training
4:30–5:20	(Company commander decides)
5:30–6:30	Dinner
6:30–9:30	(Company commander decides)
9:30–10:00	Personal time
10:00	Taps

During the nine weeks recruits master the handling of small boats as well as the seamanship required on larger craft. Recruits also spend time on the rifle range. Other topics of training sessions include:

- Officer recognition
- Drug education
- Survival
- Ranks and rates
- Nautical terms
- Communications
- Career counseling
- Small arms handling
- Seamanship
- History of the Coast Guard
- Personal hygiene
- Military customs and courtesies
- Military conducts
- Firefighting
- Human relations
- First aid

The physical part. Although the Coast Guard is usually not involved in combat, some of the day-to-day operations can be very demanding physically, as anyone who has gone out into a storm on a rescue mission can testify.

By the end of basic training each recruit is expected to perform the following:

Exercise	*Men*	*Women*
Pushups or pullups	20 4	
Flexed-arm hang		30 sec.
Bent-knee situps	40 in 2 min.	40 in 2 min.
Jump and reach or standing long jump	15 in. 6 ft. 8 in.	12 in. 5 ft. 6 in.
300-yd. shuttle run (25 yd. intervals)	65 sec.	74 sec.
1½ mi. (optional)	12 min.	13 min.

Bonuses

The Coast Guard does not offer bonuses to recruits.

Jobs

Because it has both peacetime and wartime missions, the Coast Guard offers a wide variety of jobs. Although some of them are very similar to Navy jobs, the Coast Guard does not have the Navy's large number of combat positions. It does have jobs that are not found in any of the other services, such as those related to port security, promoting boating safety, fighting pollution, keeping navigational aids operating, and enforcing customs and fishing laws.

The lack of combat jobs is especially beneficial to women. Unlike in the Navy, where women are barred from serving on combat vessels, it is possible for women in the Coast Guard to serve on or even command the largest ships. In times of war, however, women can be prevented from serving or commanding if the Coast Guard ships operate in a combat zone.

Despite the variety of jobs in this branch of the service, the small size of the Coast Guard means that the typical recruit is not offered a broad choice among them. Thus the Coast Guard does not guarantee job training as readily as do the other services.

To find out what jobs are available at any given time,

check with your local Coast Guard recruiter. Here is a partial listing of Coast Guard jobs:

- Aviation electrician's mate
- Aviation electronics technician
- Aviation machinist's mate
- Aviation structural mechanic
- Aviation survivalman
- Boatswain's mate
- Damage controlman
- Electrician's mate
- Electronics technician
- Fire control technician
- Gunner's mate
- Health services technician
- Machinery technician
- Marine science technician
- Musician
- Public affairs specialist
- Quartermaster
- Radarman
- Radioman
- Sonar technician
- Storekeeper
- Subsistence specialist
- Telephone technician
- Yeoman

11. The National Guard

U.S. Army National Guard photo

THE ARMY NATIONAL GUARD and the Air National Guard are perhaps the least understood of the nation's military forces. They look like the Army and the Air Force, operate somewhat like the Army and Air Force Reserves, and seem to get called out whenever a civil disorder or natural disaster occurs. No wonder people get confused.

The National Guard is unique in that it has a federal and a state mission. For the federal government, the Guard is a group of military men and women who can be quickly put into active duty in the event of a national emergency. They have their own equipment and are trained to fight alongside the Army or the Air Force.

For the fifty states, the Guard units represent a disciplined and equipped force that can be called out to assist during hurricanes, earthquakes, forest fires, and any other natural disasters. When civil disorders occur, such as strikes by public employees, prison breaks, or riots, the Guard can rapidly come to the aid of state or local authorities.

There are very few people who serve in the Guard full time. Most report to their units for training one weekend a month and go off for extended training fifteen days a year, usually in the summer. Members of the Guard are seldom transferred. Guard units have fought in every war, and their training exercises can take them as far as Europe or the Far East.

Currently the Army National Guard has 447,000 members operating out of 4,933 facilities. Women make up 5.5 percent of the 44,000 officers and 5.2 percent of the 403,000 enlisted people. The Air National Guard is made up of 113,000 members in 91 flying units and 244 mission support units. Twelve percent of the 13,000 officers, and 10.8 percent of the 100,000 enlisted personnel, are women. National Guard airfields and armories are located in all fifty states, the District of Columbia, Guam, Puerto Rico, and the Virgin Islands.

History

Army National Guard. The Army National Guard can trace its history back to colonial times. In fact, many units brag that they existed before the country did. In those days able-bodied men banded together in militia companies to protect towns and settlements. When a war came along they fought in it if they were so inclined. A few militias fought beside the British in the French and Indian War, and then fought against them in the American Revolution. George Washington and Thomas Jefferson at one time commanded militia regiments.

During the 1800s the militias were organized by state, rather than nationally, yet many of them took part in the War of 1812, the Mexican War, the Civil War, and the Spanish-American War.

In 1903 Congress passed laws that began to organize these state militias and to bring their training and equipment into line with that of the regular Army. The name National Guard was adopted in 1916.

All this took place just in time for World War I. The National Guard supplied seventeen divisions to the American effort, and Guard leaders noted with pride that their men fought just as valiantly as those in the regular Army.

When the war was over the Guard was dormant—until World War II appeared on the horizon. This time around the Guard provided the Army with experienced leadership, and once again the members distinguished themselves in battle. When peace came the Guard was quickly demobilized, and for a period of months it ceased to exist entirely.

The National Guard was revived, however, and the Air National Guard was split off into a separate organization. When the Korean Conflict broke out, thousands of Guardsmen were called to duty. Several units were activated during a period of international tension in 1961, and over seven thousand members saw action in Vietnam.

In peacetime the Guard has assisted in various states during floods, forest fires, and other natural disasters. It has also been called out to enforce federal court orders related to school desegregation, to perform vital services when public employees have gone on strike, and to prevent looting during riots.

One high point for the Army National Guard was the blizzard of 1978, which paralyzed cities on the East Coast.

The Guard used its personnel and equipment to clear streets and see that emergency workers got to their jobs. A low point occurred in 1970 when Ohio National Guardsmen killed four students who were a part of a large antiwar demonstration at Kent State University.

For the most part, however, the Army National Guard has served the federal government and the states as a modern version of the old militia companies—an inexpensive but highly effective force that is always ready.

Air National Guard. The first unit in the Air National Guard was established a few months after the Wright brothers convinced the Army that the newly invented airplane had a future in combat. The National Guard aviation units were not used in World War I, although many members found their way into combat through the Army's flying units. When the Guard was reorganized following the war, aviation units were again established.

By the time World War II came along the fliers in the Guard had undergone extensive training, and their presence greatly strengthened the Army Air Corps. In 1947 the name Air National Guard was adopted, and when the Korean Conflict broke out virtually all of the new organization's members saw action with the Air Force.

During the war in Vietnam, over ten thousand officers and enlisted men from the Air National Guard were called to duty. Some served side by side with the Air Force over Vietnam, while others manned bases in Japan and Korea.

As a reserve unit of the Air Force, the Air National Guard performs the same types of wartime missions as does the active Air Force, including fighter intercepting, tactical reconnaissance, air support, fighting, air refueling, and engineering. During an alert, Guard units can accompany active Air Force units or go by themselves.

In peacetime Air Guard units have participated in search missions for missing aircraft, forest fire control, and emergency missions as directed by the governors of the various states.

As the Air Force has bought new planes, the Air National

Guard has generally inherited the generation of planes that has just been replaced. In recent years, however, the Air Force has been putting new and modern aircraft in the Guard, so members are now flying F-15's, F-16's, C-141's, and C-5's.

Length of Hitch

Officers. Since Guard members normally have only thirty-nine days of active duty a year, officers generally hold down civilian jobs while they serve in the Guard.

Qualified high school graduates have three ways of becoming officers in the Army Guard. They can attend the fourteen-week Army Branch Immaterial Officer Candidate Course, the nine-week Reserve Component Officer Candidate School, or a state officer candidate school. Most state programs are taught only on weekends and take more than a year to complete. The obligation for all of these programs is three years.

Most of the officers for the Air Guard come from the Air Force. Others who wish to become officers must attend the Air National Guard Academy of Military Science near Knoxville, Tennessee. A college degree is preferred, but for some career fields it is not necessary. The course lasts for six weeks, and the obligation incurred is three years.

Enlisted. All enlisted personnel in the Army Guard and Air National Guard sign up for an eight-year hitch. The longest period they spend in uniform is during the basic training and job training they take in either the Army or the Air Force.

The rest of the eight years is divided between active Guard duty, which consists of thirty-nine days a year, and Standby Reserve, which involves no duty and pays no money.

It breaks down like this, in years:

	Basic and Job Training	*Ready Reserve*	*Standby Reserve*
Women in Army National Guard only	½–⅔	3	5
All Guardsmen	½–⅔	4	4
All Guardsmen	½–⅔	8	0

Basic Training

The Army National Guard sends its recruits to regular Army basic training. The Air National Guard sends its recruits to Air Force basic training.

Jobs

In theory the number of job possibilities in the Army National Guard is the same as in either the Air Force or the Army, but in practice this is not the case. In the National Guard you join a particular unit instead of an entire branch of the service, and your job possibilities are limited by the function of the unit and the number of skilled people it currently has.

For example, if you planned to join the 134th Refueling Wing of the Air National Guard in Knoxville, Tennessee, you would probably be unable to work as a bomb navigation systems mechanic. A unit needing someone in that specialty might be several states away, and it would probably be impractical for you to join it.

Nonetheless, the training possibilities are there. To find out what skills the units near you need, check with your local National Guard recruiter.

Bonuses

Both branches of the National Guard offer bonuses from time to time to encourage recruits to enter a job field in which there is a critical need. These bonuses can be as high as $2,000. Fields for which bonuses have been offered in the past include military police work, hospital work, and electronics.

12. How to Get Out

Photo by author

ONCE YOU HAVE TAKEN THE OATH of allegiance, getting out of the military before your period of enlistment is finished is like trying to jump off a moving train. It can be done, but the lasting effects may be much worse than staying on the train until it reaches a station.

Under the law a person cannot simply get out on request. The military has a defined set of conditions that all branches recognize as valid reasons for leaving, but the procedures can vary from service to service and from command to command. Some reasons are pretty clear-cut, such as a medical problem. If you are seriously wounded or cannot physically function, then a discharge is in order.

Other conditions can lead to a discharge. Some are administrative, as in the case of a recruit who is discovered to be sixteen years old. Other administrative conditions include disability, hardship, security, unsuitability, and a vague one called "for the convenience of the government." Depending on the record of the person involved, he or she may emerge with an honorable discharge in these cases.

Punitive discharges are another matter. These come about following conviction in a military court for a serious offense. It may be a military crime such as being absent without leave (AWOL), desertion, or a particularly bad case of disobedience. Or it may be a crime that would be a felony under civilian law, such as murder, rape, or robbery.

For someone getting out of the military, whatever the reason, the type of discharge is of paramount importance. Like a driving record, a good military discharge may not help you all that much, but a bad one can certainly make life difficult.

Discharges range from honorable down to dishonorable. An honorable discharge entitles you to all of the veterans' benefits described in Chapter 16; other discharges entitle you to fewer or none of these benefits. Since a lot of the financial advantage of serving in the military comes *after* you get out,

---

Information in this chapter was taken from several publications of the Central Committee for Conscientious Objectors as well as official military publications.

the system of graded discharge offers considerable incentive to do a good job and hang in there until your hitch is over.

Any discharge made under less than honorable conditions, or "bad paper," as it is called, will follow you like a dark cloud for the rest of your life. Most job applications ask about military service, and, although a bad discharge won't matter to some people, others will reject you out of hand for having one, no matter how well qualified you are for the job.

It is possible to have a discharge upgraded after you leave the service, but you should never count on this taking place.

GRADES OF DISCHARGE

There are five. Let's start at the top.

Honorable Discharge

If your service has been generally good to excellent in the armed forces and there are no overly bad incidents on your record, you should receive an honorable discharge. Over ninety percent of discharges are honorable.

In considering whether a person deserves this highest discharge, the military authorities examine the pattern of behavior and do not focus on the isolated incident. Homosexuals, for example, will usually get an honorable discharge unless they have confessed to or been caught in what the military calls "illegal acts." The Army's manual on discharges states that when there is doubt as to whether a person should receive an honorable or a general discharge, the doubt should be resolved in favor of the honorable separation.

General Discharge

A general discharge is also made under honorable conditions, but it is given because a person's record has not been all that good. The person involved may have been a troublemaker, but not such a pain in the neck to deserve an other-than-honorable send-off. Perhaps the person was not in long enough to establish a record. The young man who is discharged after two weeks in basic training has hardly had time to build up an excellent record. Whatever the case, a general discharge recipient is entitled to full veterans' benefits.

Other-than-Honorable Discharge

Even though this is still an administrative discharge, "other than honorable" speaks for itself. This sort of discharge is frequently handed down for misconduct, "illegal homosexual acts," security reasons, or a catchall called "for the good of the service."

An other-than-honorable discharge may cost its holder some veterans' benefits. The Veterans Administration reviews each case on an individual basis, and there is a strong likelihood that the recipient will not receive anything.

Reactions in the civilian world are hard to predict. Although this isn't a punitive discharge, the person who holds it should be prepared to do a lot of explaining.

Bad-Conduct Discharge

Now we have passed into the realm of punitive discharges—ones that clearly label your military experience as unsatisfactory. A bad-conduct discharge is a result of a conviction in a court-martial and is roughly equivalent to a misdemeanor or minor criminal conviction in civilian life.

Again, the V.A. decides on a case-by-case basis if the holder gets any benefits, but with a bad-conduct discharge the deck is stacked against you.

Dishonorable Discharge

This is the worst discharge a military person can receive. It indicates that you have been convicted of a serious crime—one that might be considered a felony outside the armed forces.

This ultimate "bad paper" strips the holder of most of the veterans' benefits, and in some states he or she will even lose the right to vote. Getting a job with a dishonorable discharge can be extremely difficult.

SPECIAL CONDITIONS FOR DISCHARGE

The five grades of discharge are clearly established, but the conditions that result in each one aren't. Discharges are a tool for the military, and different branches and different commanders use them in varying ways.

The discharge system first of all encourages military personnel to work hard and do a good job. The promise of benefits after leaving the service is enough to motivate people, and the threat of a bad discharge and the consequences that follow help to keep most military personnel in line.

The second purpose of the discharge system is to make it easy to get rid of people who cannot or will not function in a military setting. Slight physical or mental disorders could prove life-threatening in a combat situation, both to the person with the disorder and those around.

As for those who will not function, the armed services tolerate a narrow range of behavior, and characteristics that might be considered acceptable or even admirable in civilian life—such as nonconformity or spontaneity—do not always work in a military setting. The armed forces operate best when unimpeded, and those who go against the flow may find themselves heading for a discharge.

You cannot apply for discharge under most of the conditions that lead to it. There is no form, for example, to tell your commander that you have a behavioral disorder. The military assumes that such a condition will be obvious to the authorities.

There are some situations, however, where an enlisted person can apply for special consideration by the military. They include, but are not limited to, hardship, dependency, erroneous enlistment, and conscientious objector status.

Dependency

Sue is ten years older than her brother. She joins the Air Force, and, while stationed in Greenland, learns that both her parents have been killed in an auto accident. She is the only close relative her twelve-year-old brother has.

Sue can apply for a dependency discharge. An application and supporting documents must generally be submitted to the commanding officer, who forwards them to the proper authorities. Sometimes the applicant is interviewed for further information, but often the documents that accompany the application are enough.

Usually a person who applies for discharge due to dependency receives an honorable or a general discharge, depending on his or her record.

Hardship

While Roger is serving in the Navy his unemployed father falls off a ladder and breaks his back. The resulting medical bills force the family, of which Roger is the only child, into near bankruptcy. Even the money that Roger sends home is insufficient to pull the family through.

Roger can apply for a hardship discharge. As with Sue's dependency situation, an application must be filled out and accompanied by documentation that supports the claim. This sort of discharge is more difficult to achieve with the advent of social service funds such as Medicare, Medicaid, food stamps, and so on.

As with the dependency situation, the person who is approved for hardship status generally receives an honorable or general discharge.

Erroneous Enlistment

Hernando joins the Navy to be an air traffic controller. He asks his classifier to make sure he is qualified for the training before he signs up. He is assured in writing that he is qualified for air traffic control training, and he enlists.

However, with 20/200 corrected vision, Hernando is not qualified for air traffic control school, and when it is learned that he has relatives in Cuba it becomes obvious that he probably would never get the security clearance to become a military air traffic controller.

Hernando is a good candidate for an erroneous enlistment discharge. Anyone receiving a written guarantee of training is entitled to get it or be discharged. (This is a good reason to get all of the promises made to you at the processing station in writing. If they were only spoken, Hernando would have a very hard time making his case.)

A person in such a situation should prepare a statement requesting discharge and explaining the problem. As with any application for discharge, any and all supporting documents, such as the recruiting contract, should be included. (This is a good reason to always keep a copy of anything you sign.)

If Hernando is successful in his quest he will receive an honorable discharge.

Conscientious Objector

When you enlist in the armed forces, one of the questions on the form that you sign asks: "Do you now have, or have you ever had, a firm, fixed, and sincere objection to participation in war of any form or to the bearing of arms because of religious training or belief?" If you answer yes this will not necessarily prevent you from applying for enlistment, but you are unlikely to be accepted into the military.

Even after you have answered the question with a no, military officials recognize that while in uniform you may develop a conscientious objection to war or the bearing of arms. They don't like it, but they recognize it.

A person may become a conscientious objector (C.O.) in several ways. Some become C.O.'s after experiencing combat or being ordered into combat. The young woman who entered the military to gain a job skill or money for college may realize with a shock that the B-52 bomber she is repairing carries nuclear weapons. Or it may suddenly dawn on the recruit on the Marine rifle range that he is being trained to shoot at human beings.

However you become a C.O., you can get an honorable discharge from the military. To do so you must convince the authorities that you came to your C.O. beliefs after entering the military. You do not have to be a member of a church—you don't even have to believe in God—nor do you have to renounce the use of force in all situations.

All of the services have procedures for filing for a C.O. discharge. It is not complicated, but it takes time and a lot of perseverance. You must prepare to undergo isolation and perhaps some abuse from those around you. But it can be done. Since 1970, according to an organization that assists persons applying for C.O. status, almost ten thousand military members have applied for C.O. discharges and over six thousand have been granted.

If you join the military and have a change of heart, talk to a civilian counselor or attorney who is familiar with military matters. An ordinary attorney will probably not know all the regulations and practices. A good counselor can help you think through your position—he or she may have even been in your shoes at one time—and can advise you on the best way to proceed.

MILITARY COUNSELING

You don't have to be in uniform to talk to a military counselor. If you are unsure about joining or have nagging doubts about your role in the armed forces, talk to a counselor *before* you enlist.

You can find military counseling by looking in the phone book, by talking with local religious leaders, or by calling one of the agencies listed as follows. Here are several sources of help:

☐ Central Committee for Conscientious Objectors (CCCO), An Agency for Military and Draft Counseling, 2208 South Street, Philadelphia, Pennsylvania 19146. (215) 545-4626.

☐ CCCO–Western Region, P.O. Box 42249, San Francisco, California 94142. (415) 552-6433.

☐ Midwest Committee for Military Counseling, Inc., 59 East Van Buren Street, Suite 809, Chicago, Illinois 60605. (312) 939-3349.

□ Military Law Task Force, National Lawyers Guild, 1168 Union Street, No. 201, San Diego, California 92101. (619) 233-1701.

□ American Friends Service Committee (AFSC) Headquarters, 1501 Cherry Street, Philadelphia, Pennsylvania 19102. (215) 241-7000.

AFSC Regional Offices:

92 Piedmont Avenue, N.E., Atlanta, Georgia 30303. (404) 586-0460.

1022 West Sixth Street, Austin, Texas 78703. (512) 474-2399.

317 East Twenty-fifth Street, Baltimore, Maryland 21218. (301) 366-7200.

2161 Massachusetts Avenue, Cambridge, Massachusetts 02140. (617) 661-6130.

Suite 1400, 59 East Van Buren Street, Chicago, Illinois 60605. (312) 427-2533.

915 Salem Street, Dayton, Ohio 45406. (513) 278-4225.

4211 Grand Avenue, Des Moines, Iowa 50312. (515) 274-4851.

15 Rutherford Place, New York, New York 10003. (212) 598-0950.

980 North Fair Oaks Avenue, Pasadena, California 91103. (818) 791-1978.

2160 Lake Street, San Francisco, California 94121. (415) 752-7766.

814 Northeast Fortieth Street, Seattle, Washington 98105. (206) 632-0500.

13. Women in the Military

U.S. Marine Corps photo

ONE OF THE BIGGEST changes to hit the military has been the active recruitment of females. Women have served in uniform for over fifty years, but until 1973 their activities were for the most part confined to traditionally feminine jobs such as nursing or clerical work. With the end of the draft, however, there was a great need for what had formerly been called manpower, and women moved into many new fields. Now women can be found serving as F-16 crew chiefs in the Air Force and electronics specialists in the Navy, and stationed beside men in Europe. They make up 10 percent of the officers and enlisted people in uniform today.

As you can see from the table, in every branch of the service men still vastly outnumber women. But women are better represented in the U.S. military than in the armed forces of any other country. Israel is often cited as a country with a strong female presence in the armed forces, yet women make up less than 3 percent of the Israeli military. The U.S.S.R.'s forces include women, but they make up less than 1 percent of the active military personnel, they are largely confined to medical and clerical duties, and not one is a general.[1]

Despite the great strides that women have made in the American military since 1973, neither they nor the service has entirely adapted to their expanded role. Women in uniform still face special problems.

Percentage of Women in the Service

Branch	*Officers*	*Enlisted*
Army	10.0	10.3
Navy	10.0	9.1
Air Force	11.2	12.1
Marines	3.3	5.4
Coast Guard	2.8	7.6
National Guard	5.5	5.2

1. Michael Levin, "Women as Soldiers—The Record So Far," *Public Interest* 76 (1984): 33.

History

The role of women in combat has varied drastically over the past three hundred years. Recent wars have conditioned Americans not to associate women with combat. Females have on occasion found themselves in action—when hospitals have come under attack, for example—but for the most part they have been kept away from fighting.

This was not always the case. In colonial days, when isolated settlements were attacked by Indians, anyone who could fire a rifle did so, and women as well as men took up positions in forts and stockades.

This early tradition of women fighting alongside men gradually evolved into one of women serving as camp followers. In the ragtag armies of the revolutionary war women came along to work as nurses, cooks, and laundresses. But more than a few took part in the fighting. The most famous of these is Molly Pitcher, who filled in for her wounded husband on a cannon crew and eventually was welcomed into the army by George Washington.

Several women took part in battles, some on an emergency basis like Molly Pitcher and others who volunteered for full-time duty. Some went so far as to dress like men in order to join. This pattern continued through the War of 1812 and even into the Civil War, when women took part in combat on both sides.

Although it marked the end of women's participation in combat, the Civil War also marked the beginning of efforts to organize the camp followers. Abraham Lincoln authorized Clara Barton to create a regiment of women nurses. In 1901 Congress formally established the Army Nurse Corps, and the Navy Nurse Corps followed in 1908. Despite the fact that they were part of the armed forces, these nurses could not hold rank, nor were they entitled to the same benefits as men.

In World War I women began to move out of the hospitals and into administrative posts to free men for combat. This time around they were given rank and were officially discharged when the war was over. Over one hundred thirty thousand women served in this capacity in World War I.

The role of women in World War II was considerably expanded with the creation of the Women's Army Corps (WAC), the Navy's Women Accepted for Volunteer Emergency Service (WAVES), the Women Airforce Service Pilots (WASP), and the Marine Corps and Coast Guard Women's Reserve. Two hundred sixty-five thousand women were in uniform during the war, and their duties included flying new aircraft to the front, instructing men in basic training, and the usual clerking and nursing.

Although an effort was made to keep women out of combat, many were present where fighting took place. Nurses were taken prisoner by the Japanese, and other women—like their frontier ancestors—picked up guns and fought alongside the men when things got desperate. On the home front, civilian women proved that they could handle difficult factory jobs once limited to men.

By 1948 the authorities were convinced that there was a permanent role for females in the armed forces, and women were allowed to enlist in active units and the reserves. Limits were placed on the total number of women who could join up and the rank they could attain, and they were barred by law from flying combat aircraft or serving on naval combat vessels. Oddly enough, the law did not prevent them from fighting on the ground, although an internal Army policy accomplished that.

Women served with distinction in Korea, but it wasn't until Vietnam that their numbers in the military rose significantly. The ceiling on the number of women in uniform was lifted, and close to one hundred ninety-three thousand women were on duty during those years. Approximately seventy-five hundred actually went to Vietnam, mostly as nurses.

Then came the end of the draft in 1973. The country was disenchanted with military service, primarily because of Vietnam, and the various branches of the armed forces had to scramble to meet enlistment quotas. For the first time they began to recruit women in nontraditional fields. There were approximately fifty-five thousand females on active duty in 1973, and this number rose to 111,753 in 1976 and 184,651 in 1981. Currently there are approximately two hundred ten thousand women in uniform.

With their numbers increasing so quickly, pressure was brought to admit women into the service academies. Traditionalists in every branch fought the move, and male cadets huffed and puffed about how women would never be able to take the physical and psychological pressures that mark academy life.

Congress disagreed, however, and in 1976 women were admitted to the academies. These female cadets were subjected to intense scrutiny by the media, and whenever one would drop out the papers would carry a story. But most of them prevailed; they graduated and were commissioned as officers. Although there is still grumbling in some quarters about women in the academies, most people consider the change a success.

With women so firmly a part of the nation's armed forces, a growing number of voices inside and outside the military are arguing that women should take part in combat. An equally vociferous group insists this should never take place. If the restrictions against combat are ever dropped, the women in the American military will take their place alongside their male counterparts much as their pioneer ancestors did. They will have come full circle at last.

Things Women Should Consider

Aside from the usual questions confronting someone who is considering military service—what branch to join, whether to become an officer or enlisted person and the like—women have additional matters to consider. Society does not always prepare women for the performance expected in a military setting. Once in uniform, females may encounter problems caused by physical differences, the military institutions themselves, and the men in uniform.

Just like their male counterparts, some women go through

the military with a minimum of bother, while others are miserable from day one. Before you sign the enlistment papers, here are a few things you should think about.

Reactions in the civilian world. The woman who announces to her family and friends that she is enlisting in the armed forces will get a reaction different from that a man would get.

"I didn't know you were that desperate" is a comment often heard. Some people see the military as an employer of last resort and assume that a woman is at the end of her rope when she joins.

Others imagine that the only reason a woman would join the military is to find a husband. A smaller group will accuse an enlistee—though not to her face—of being promiscuous or of having no self-esteem. And a minority may mutter comments about her being a lesbian.

Close friends or others who would snort at the above accusations will often question how a young woman could involve herself with anything involving war and destruction. They see this as an exclusively male activity and wonder why a woman, who is trained by society to be a nurturer, would want any part of it.

You can't prevent someone from making negative remarks about whatever you do, whether it is joining a religious order or flying an Army helicopter. That's part of life. You can, however, talk to people about your own feelings regarding military service. Present or former servicewomen can give you some good advice, or at least explain how they resolved the issue. If nothing else, they may supply you with a quick line to squelch those folks who inevitably sniff that "the military is no place for a girl."

Physical differences. When women first started filling nontraditional roles in the military, they found that the uniforms and boots and pieces of equipment they were supposed to use were not suited for women's bodies. These problems have largely been worked out by now, but the physical differences between men and women are still very evident in some quarters.

A 1982 Army study found that women's upper-body strength is only 58 percent that of men. Furthermore, due to their lower muscle-to-fat ratio, women rank 20 to 40 percent below men of equal weight on strength tests.[2] This has caused problems in training and sometimes on the job. At boot camp, women now perform different exercises and meet different requirements. Instead of chin-ups they do a "bent-arm hang."

Changing the requirements during basic training was easily done, but requirements on the job are something else. The weight of a cylinder of acetylene for use with a cutting torch, or the cast iron head of a diesel engine, remains the same for women as for men. In some military jobs women may be accused by their coworkers of not pulling their share of the load.

Finally, men are not always understanding and sensitive regarding the problems some women suffer with menstrual periods. Trying to explain severe cramps to a male drill instructor or commanding officer may be embarrassing and frustrating.

Benefits from a few years in uniform. Aside from the benefits that are available for everyone in the service, a military stint can be especially helpful for women starting their careers. In the civilian world women are often paid less than men for the same kinds of jobs. Not so in the armed forces. If you are a staff sergeant, then you get the same base pay and allowances as any other staff sergeant in your situation. One three-year study revealed that young women in uniform made more money on the average than did their civilian counterparts.[3]

Another benefit of entering the service can be the training you receive. A woman can enter a trade in the military that might be much harder to get into in the civilian world. If you participate in one of the apprenticeship programs, you can come out of the service qualified to work in a high-paying

2. Levin, "Women as Soldiers," 34.
3. Lois B. Defleur and Rebecca L. Warner, "Socioeconomic and Social-Psychological Effects of Military Service on Women," *International Journal of Sociology and Social Policy* 3 (1985): 197.

field. And your veteran status may help you in getting certain Civil Service or government jobs.

The biggest advantage of a short stay in the military may be the chance to learn the ropes in a competitive, male-dominated organization. Women who can successfully operate in such an environment will find things easier when they go to work in the civilian world.

You may learn to perform fine in a male setting, but one obstacle will remain no matter how good you prove yourself. The law against flying combat aircraft and serving on Navy combat vessels, plus the Army's policy of keeping women out of combat, will have a considerable impact on your military experience.

For one thing, it will mean that you will choose from a smaller selection of jobs than will a man. Even if you are admitted to a certain job classification, you might be unable to perform your duties because you are not allowed to go where the work is done. For example, a female Navy computer repairer cannot ship out on an aircraft carrier; it is a combat vessel.

Curiously enough, although the Army says women may not serve in combat, females are stationed throughout West Germany. If Warsaw Pact troops ever come pouring across the border from the east, there will be little if any time to evacuate the women.

Making a career in the military. There is certainly room at the top for female officers. In the Marines, for example, only 647 female officers are on active duty, compared to 19,536 male officers. Even though the armed forces have been recruiting women since 1973, women were first admitted to the service academies only in 1976. The first year 357 were admitted, and when the cadets threw their hats into the air four years later at graduation, 214 women were left; 40 percent of the women made it to graduation, as compared with 37 percent of their male counterparts. Having now served for five years, that first crop of women have their first opportunity to call it quits. A recent *New York Times* article estimated that 30 percent of them will hang up their uniforms for good; this is approximately the same percentage of men who resign at five years. That leaves about one hundred fifty. Even counting women who come in by other means, it will take a long time to increase the number of high-ranking female officers.[4]

The biggest problem that career military women face is the ban on combat. Whereas this can be a small hindrance to those who are in for a few years, those who have their sights on the highest levels will have to work very hard to compensate for a lack of combat experience. A high-ranking Navy officer, for example, must at some time in his or her career command a ship. Men can be selected for aircraft carriers, battleships, frigates, and all the other exciting vessels. Women are left with mine-sweepers, tankers, and other vessels that, though they may serve an important function, are very unlikely to ever engage in battle.

If a war or skirmish breaks out, the officers in a combat area have an opportunity to demonstrate their leadership under fire. If they do well, they better their chances for promotion. This path of advancement is closed to women, making their assumption of the highest ranks much less likely. The situation may change in the future, but you should make your plans assuming it won't.

Sexual harassment. According to the Department of Defense's definition, sexual harassment can range from obscene comments or gestures to sexual contact. It is perhaps the worst problem a woman will face in the armed forces.

The situation is not so much an institutional as an individual problem. From the Secretary of Defense on down, each of the armed forces deplores sexual harassment and has issued countless memorandums on the subject. If a clear-cut case comes to their attention, they usually move quickly to discipline the guilty party.

Sexual harassment can take many forms. The man who constantly uses obscene language or tells sexual jokes in your

4. Esther B. Fein, "The Choice," *New York Times Magazine*, May 5, 1985.

presence is guilty of sexual harassment. The person who constantly asks for a date even though you have made it clear you aren't interested is guilty of sexual harassment. And the officer who suggests that going to bed with him is the only way you will get promoted is certainly guilty of sexual harassment.

Witnessing or being a victim of sexual harassment is one thing, but getting anything done about it is something else. The problem stems mainly from the rank system—the chain of command. The person commanding you has considerable power over you. He can influence your advancement, your assignments, and even small matters such as your leaves and liberties.

Sexual harassment can easily take place given this power of one person over another. And the arrangement that makes it easy to happen makes it difficult to punish. If you want to make a complaint about sexual harassment you have to go to the person above you. And if he is the culprit, you have to go to the officer above him. Jumping the chain of command is rarely done in the military, and making this move takes a lot of courage.

Assuming you accuse the officer above you of sexual harassment, how can you prove it? Most sexual demands are not made in public, and it may simply be your word against his. If the superior officer decides to drop the matter or rules in favor of the offender, you are still under the accused's command—and may be in hot water. Women sometimes fear they are in physical danger after making an accusation of sexual harassment.

The outright sexual invitation or threat is an example of serious harassment, but sometimes the little things are the most pervasive and irritating. The *Stars and Stripes,* a newspaper for military personnel, ran an article in 1982 detailing cases of sexual harassment in American military communities in Europe. One woman stopped eating in the mess hall because of the constant leers and comments. She described it as "an unpleasant, anxiety-producing experience."[5]

---

5. *Stars and Stripes,* July 23, 1982.

What should women do? Another *Stars and Stripes* article suggests confronting the offender and asking him to stop. If this doesn't work, the victim should report the incident to the next person in the chain of command, an equal opportunity officer, or a women's advocacy officer. The story ends with a promising quote from an Army researcher: "There's a growing number of commanders out there who want to do something about the sexual harassment problem. But there's nothing they can do if women don't report it."[6]

Some women can shrug off sexual harassment, while others are driven to leave the military because of it. You should consider how you would react to this unfortunate aspect of military service. It shows little sign of going away soon.

Family life. Family life in the armed forces is certainly not just a woman's concern, but matters such as a two-career marriage and pregnancy are of particular concern to women considering the military.

A family begins with a marriage. There is a fair possibility you may decide to marry someone in your branch of the service—over forty-five thousand couples have done so. Just as in civilian marriages, military couples have to make decisions about which career paths to follow. Staying together can be difficult if your spouse has a combat job and gets stationed in areas where you cannot go. Making sure you stay together can limit job possibilities and affect advancement—yours and his.

Marriage to someone in another branch of the service is possible, but that really makes things complicated. And if you marry a civilian, he may not appreciate being dragged from base to base and job to job.

Children complicate matters further. The armed forces used to regard pregnancy as grounds for automatic discharge, but this is no longer the case. In fact, a 1984 article claimed that 10 to 15 percent of the women in the Army are pregnant at any given time.[7] If you become pregnant while in uniform,

6. *Stars and Stripes,* July 21, 1982.
7. Levin, "Women as Soldiers," 38.

you can request a discharge and it will be granted to you. But if you want to stay in the military after having the child, you will probably be guaranteed no special consideration in assignments or duty stations based solely on the fact that you are a mother. Unless medical conditions dictate otherwise, you are expected to be back in the saddle six weeks after delivery.

Women who want to combine a military career and a family cannot drop out or go to half-time for a few years in order to spend time with the children, as their civilian counterparts often do. Furthermore, the irregular hours for military personnel—men and women—make arranging consistent child care very difficult. And when the inevitable sicknesses hit the child, and each parent has a position of responsibility, they both face the question that increasingly confronts young parents: Who stays home? Even if you find a perfect child care situation, the frequent moves mean that you will have to start all over again in two to three years.

As children grow older, a military life can mean changing schools as often as every year, constantly making and leaving friends, and perhaps showing deference to the children of those who outrank you. Thousands of children have been raised as military dependents; some of them have thrived or at least didn't suffer from the experience, whereas others have truly earned the title "military brat."

Outlook

For women, the military is a two-edged sword. On one side it offers job training, money for education, equal pay, and a chance for a meaningful career. On the other it imposes obstacles simply because of your sex, exposes you to possible sexual harassment, and can make raising a family very difficult.

It's important to remember that the civilian world is not all sweetness and light, either. Sexism is not confined to men in uniform, discrimination against women is common in business and industry, and combining family and a meaningful career is never easy anywhere.

Before entering the service, weigh your options carefully. If possible you should talk with military women—both enlisted and officers—for a better sense of what to expect. The things they will tell you will be worth ten times what you'll hear from a male recruiter.

14. Blacks in the Military

U.S. Marine Corps photo

BLACKS AND THE MILITARY constitute a study in contrasts. On one hand the armed forces, through their training and benefits, offer some of the best ways for young people to make the most of themselves. On the other hand they can subject an individual to the last thing he or she needs more of: racism.

Even the history of blacks in the armed forces has shown vast swings. For decades the numbers of blacks were purposely kept low, and they were confined to the most menial tasks. Then, in a complete turnaround, the Department of Defense became the first major American institution to be integrated.

Charles C. Moskos, a prominent sociologist who has studied the military, once stated, "I'd say that racism—the American dilemma—is *the* problem of all the armed services, wherever the troops are." Yet he added, "I believe the military has gone further in attacking racism than any other institution our society has."[1] Ten years later he published an article entitled "Success Story: Blacks in the Army."[2]

History

The role of blacks in American military history roughly parallels that of women. Blacks went from fighting alongside whites to serving as camp followers, then working in support positions before finally taking their place again in all aspects of combat.

Most histories of blacks in combat begin with Crispus Attucks, the first American to die in the infamous Boston Massacre of 1770. Black citizen-soldiers fought beside their white counterparts at Lexington and Concord. They took part in nearly every engagement of war, sometimes on both sides. Not surprisingly, considering the slaveholding policies of the colonials, more blacks fought for the British than for the revolutionaries.

---

1. Bruce Bliven, Jr., *Volunteers, One and All* (New York: Reader's Digest Press, 1976), 120–21.
2. Charles C. Moskos, "Success Story: Blacks in the Army," *Atlantic*, May 1986, 64–72.

After the Revolution the young United States took steps to relegate blacks—like women—to the status of camp followers. Despite this, close to a thousand blacks saw action in the War of 1812. Most were excluded from the armed service, however, and those permitted to join served officially as cooks or servants.

This state of affairs remained until the middle of the Civil War, when the North realized it had a valuable untapped resource. Several all-black companies were established, and after the Emancipation Proclamation of 1863 volunteers readily filled the special regiments. Over seven thousand blacks were commissioned as officers, although most of these were doctors or chaplains. When the war was almost over even the South saw fit to allow blacks to join up and fight. (There is no record, however, of many availing themselves of this privilege).

Having proven themselves in action, black units were maintained and sent to the West where they fought with the Indians. They became so proficient at cavalry tactics that two black units accompanied Teddy Roosevelt in the famous attack on San Juan Hill in Cuba during the Spanish-American War.

Despite their demonstrated willingness and readiness to fight for their country, "World War I saw official armed forces policy toward blacks at its worst—rigid segregation, partial exclusion, and open and virulent racism on the part of numerous white commanders of black units."[3] So writes one historian. Combat-trained black troops were put to work as laborers or cooks. Even during this low in minority-military relations, however, one black unit—the 370th Regiment, one of the few with black commanders—performed superbly and was highly decorated.

But after the war it was back to segregation as usual. Black soldiers in the Army could not serve in the Air Corps or the artillery, engineers, signal, or tank corps; and the Navy and Marines kept their blacks tucked away in the kitchen.

---

3. Richard J. Stillman II, "Black Participation in the Armed Forces," in *Black American Reference Book*, Mabel Smythe, ed. (Englewood Cliffs, N.J.: Prentice Hall, 1976), 894. Much of the material in this chapter comes from Stillman's excellent work.

World War II marked the beginning of the change. Much as in the Civil War, blacks were underused at first. Like the women in uniform, blacks found themselves in support positions instead of combat. They unloaded ships, drove trucks, and took their usual positions in kitchens and servant quarters.

When manpower shortages occurred late in the war, blacks were taken out of the support positions and sent to the front. A new Naval Secretary, James Forrestal, integrated basic and advanced training, put 10-percent-black crews on certain ships, and commissioned the first black naval officers.

Many blacks who participated in the fighting distinguished themselves; several units were highly decorated, and generals such as George Patton spoke highly of the black troops under their command.

But there were still segregated units, and their members were forced to endure separate officers' clubs, noncommissioned officers' clubs, and recreation facilities. The Red Cross even established a separate blood bank for blacks. Only this time the blacks didn't take it sitting down. When they were treated unfairly by white officers they complained, and sometimes they went further. Riots broke out at Army bases in Hawaii, Georgia, and Louisiana, and the Navy experienced serious racial clashes at ports in San Francisco and Guam.

The change was long overdue, and in the late forties the segregational barriers began to crumble. President Harry Truman signed an executive order that established equal treatment and opportunity for all persons in the armed services. The newly created Air Force set the pace by complying with the order first; the Navy made impressive gains, while the Army held onto its segregated units.

It took the war in Korea to bring the Army into line. Blacks enlisted in large numbers, and it was impossible to put all of them into the special units. By the end of the war the armed forces were entirely integrated—the first major institution in the country to become so.

Despite the advances made on an institutional level, blacks in the armed forces still faced problems during the Vietnam years. Unlike in earlier wars in which they had initially been kept away from combat, blacks, many of whom had been drafted against their will, fought in large numbers in Vietnam. This was a matter of contention nationwide; many people felt black troops were bearing the brunt of the fighting while whites were sitting at home with student deferments. Actually, however, only 12.1 percent of the Americans who died in Vietnam were black—about the same percentage of blacks in the American population.[4]

Other issues surfaced during and immediately after the Vietnam period. The rise of black militancy and a corresponding unwillingness to tolerate racism led to barracks battles at military installations and on ships. A few white troops organized Ku Klux Klan chapters, and tensions ran high. One German employee of the Army in Europe offered this analysis of the early seventies: "In the volunteer Army you are recruiting the best of the blacks and the worst of the whites."[5] This was clearly an exaggeration, but it contained a kernel of truth.

The Situation Today

A black person entering the service today will see more people of color in positions of authority than in virtually any other area of our society. Because the military is such a controlled society, a lot of the problems that plague minorities in civilian life have vanished here. The pay is the same regardless of race. So is the housing. So is the access to medical care. And so are the schools for the children of servicemen and -women.

Blacks are more heavily concentrated in the Army and in the Marines than in the other branches of the service. In the lowest ranks, 27 percent of the soldiers in the Army are black. Blacks make up 19 percent of the Marine Corps' lowest ranks, 15 percent of the Air Force's, and 12 percent of the Navy's.

The highest percentages of blacks are found halfway up

---

4. Moskos, "Success Story," 66.
5. Brent Scowcroft, ed., *Military Service in the United States* (Englewood Cliffs, N.J.: Prentice Hall, 1982), 133.

the enlisted ladder, reflecting the situation five to ten years ago when many more blacks than whites entered the service. Thirty-eight percent of the Army's sergeants are black, along with 25 percent of the Marines'. Blacks make up 18 percent of the Air Force's sergeants and 13.5 percent of the Navy's second-class petty officers (whose rank is equivalent to sergeant).

At the highest enlisted rank—sergeant major—blacks are represented as follows: The Army has 30 percent blacks, the Marines 17.5 percent, the Air Force 11 percent, and the Navy 6 percent.[6]

In the officer ranks the percentages diminish even more. Blacks are concentrated in the lowest three levels. Of all officers, the Army once more leads with 10 percent. The Marines and Air Force have half that percentage, and the Navy has only a third of it.[7]

What do all these comparisons mean? It appears that blacks are making it in the enlisted ranks; their strong representation at the middle of the ladder should soon bring about equally strong representation at the top. But blacks have a long way to go before they will raise their numbers to equivalent levels in the officer corps. The competition is greater there, and blacks who qualify for entrance to a service academy or ROTC may be heading toward lucrative civilian opportunities instead of the military.

Why are blacks concentrated in the Army and Marines? These branches of the service, compared with the Air Force and the Navy, have many more positions that require no technical skills. Although more black men and women enlist with high school diplomas than do their white counterparts, the whites still average higher scores on the ASVAB and other standardized tests. For a person with low test scores, the Army and Marines offer more job opportunities than do the other branches.

According to one study of the Army, "blacks are more

---

6. Richard Halloran, "Women, Blacks, Spouses Transform the Military," *New York Times*, August 25, 1986.
7. Moskos, "Success Story," 66.

likely than whites to be assigned to 'support' branches of the service. They make up 50 percent of those in supply, 46 percent of those in food service, and 44 percent of those in general clerical work. Blacks are less likely than whites to be found in highly technical fields, such as signal intelligence, cryptography, and electronic warfare. And in combat specialties—the guts of the Army—black participation has been declining. . . . Although blacks are overrepresented in combat specialties relative to their numbers in American society, they are considerably underrepresented relative to their numbers in the U.S. Army. Despite popular perceptions, black males are not being tracked into combat units."[8]

Though blacks have done well in the service, some things are harder for them than whites. Racism in the armed forces is still present, although nowhere near as blatant as it used to be. Since promotions are based on written assessments of an individual, it's very easy for a superior to hold someone back with faint praise instead of an accurate assessment of his or her ability.

The military regularly commissions opinion surveys to discover what is on the minds of the enlisted personnel and the officers. The results are encouraging. Blacks are more likely than whites to detect racism, not surprisingly, yet blacks are more content with their work.[9]

Things Blacks Should Consider

Military service makes unique demands on a person. You are expected to obey orders without question, to take abuse and not reply, and to show respect and follow the commands of any person who outranks you. And you can't quit.

Furthermore, in a military setting you may be thrown into close contact with people you would normally not associate with. You may have to work or even live with a person whose whole outlook on life is opposed to your own. Despite the human relations courses, despite the attention paid to racial

8. Moskos, "Success Story," 66.
9. Moskos, "Success Story," 70.

matters, and despite the countless letters and memorandums circulated on these matters, there are still out-and-out racists in uniform. They may be in your company. Worse yet, they may be in charge of your company. You have to learn to deal with these people without detracting from your goals in the military. It isn't always easy. Here are some questions you should ask yourself.

Racism: can I handle it? If you grew up in an all-black neighborhood and attended high school with lots of your friends, you may never have encountered some of the more ugly forms of racism. You may never have had a white person call you "nigger" to your face. How will you react if this happens? Will you get into a fight? Report the incident? Or try to get back at the person?

If a white person gets a job you were qualified for or a promotion you should have had, what will you do? Will you let it eat away at your insides, gradually filling you with hate? Will you throw up your hands and give up, sullenly marking time until you are discharged? Or will you work harder so you will have a better chance of getting the job or the promotion next time?

In thinking about joining the armed forces, you have to look at yourself as well as the enlistment options. You may come to the conclusion that you would not operate well in a military setting; it's a lot better to recognize this *before* you get in.

What if I am assigned to a tough duty station? Most people assume that when they join the service there will always be some place to go on the weekends, people to meet—particularly of the opposite sex—and opportunities for good times. The brochures in recruiting offices often picture racially mixed groups having a party on a beach, sitting in a nice restaurant, or skiing down a mountainside. This isn't always the way it is. You may be assigned to a base near a town in the boondocks where there are no nonwhites. The only place to go on Saturday night may be the Redneck Dew Drop Inn. Or you may be overseas where no one off the base speaks English and those on base want nothing to do with you.

In considering this possibility, you should examine the various branches of the service and find out where you might be stationed. In selecting your job training, inquire of the recruiter where such jobs are most often done. Ask if there is a base-of-choice or country-of-choice option.

Can I get a better deal in the military than as a civilian? This question is the bottom line, and there's no way of definitely answering it until you're in. And then it's too late to back out.

The military still poses a lot of problems for minorities. Racism still exists. Blacks are imprisoned in military stockades at a higher rate than whites. There is a greater proportion of white officers, and blacks are more often found performing a menial job than learning a technical skill. Minorities are more likely than whites to receive a dishonorable discharge. Some black veterans are extremely bitter about their treatment in the armed forces, and quite a few feel they are worse off for the experience.

Yet the potential for bettering yourself is there. Despite the racism and raw deals—which aren't unheard of in civilian life—you can go into the armed forces with nothing and come out with a useful skill or money that will help you go to school and make something of yourself. For many people the military was the first break on the way to success. If you can put up with the hassles and the unpleasant people, if you can work in spite of the adversity, the service can be the best thing that can happen to you.

In both officer and enlisted ranks, the military offers blacks an important benefit: it teaches how to succeed in a system dominated by white males. Like it or not, this is the kind of system most veterans will enter sooner or later, and if you've done well in uniform, you'll have a better chance of making it on the outside.

The smartest thing to do before signing the enlistment papers is to talk to a minority veteran or someone who is still in uniform. Ask them the hard questions; get them to tell you about their good and bad experiences. Don't talk only to recruiters—even black ones. Things must have gone well for them, or they wouldn't be behind that desk.

If you decide to go into the service, at least you'll enlist with a clear idea of what to expect—good things and bad.

15. Educational and Post-Service Benefits

Official U.S. Navy photograph

PERHAPS THE MOST SOUGHT-AFTER benefits of military service are educational. Money for education is available before, during, and after active duty. Once you are back in civilian clothes you are eligible for other benefits as well: veterans are sometimes favored in hiring decisions, and you may be entitled to a pension, a loan guarantee, and free medical care.

EDUCATION IN THE ARMED FORCES

By and large the military authorities have always felt that educated military personnel make up a superior force. Such people can be trained more quickly, can operate complicated weapons more effectively, and, when properly motivated, can fight better.

For these reasons the armed forces have long encouraged their troops to educate themselves. The General Education Development Test (GED)—the high school equivalency test—originated in the military, and millions of classroom hours have been logged under military supervision.

Since the end of the draft, however, the attention to education has reached an even higher priority. There are still programs through which military personnel can get their high school diplomas, but the emphasis has shifted toward college and graduate education.

All this emphasis on education serves three important functions for the armed forces. First, it brings in better people. Those with little schooling and few skills have always gravitated to the military, but increased educational opportunities lure the best recruits. These people train quickly, stay out of trouble, and operate sophisticated (and costly) equipment well.

Second, money for education helps keep these good people in uniform. Officers who are sent to graduate or professional schools incur longer obligations to serve. The more specialized their training, the more useful they become to the military. They stand better chances of being promoted and less often resign their commissions for civilian jobs.

Finally, leaders in the armed forces feel that money spent

for education benefits the military as a whole. Sending men and women into the civilian world to learn the latest in technical and managerial techniques can help the military perform more efficiently and effectively.

What's in it for me? Most people aren't interested in getting an education just so the military can perform more efficiently and effectively; they want to know what is in it for them.

There are several advantages to getting your education through the military. The first is saving money. The average student at a four-year public college graduates owing close to eight thousand dollars; those who go to private schools often owe even more. The military offers scholarships and programs that can lower or eliminate these costs. And whereas most civilian scholarships go to those who have high grades, the military has money for those who average B or C grades as well as for high achievers. Finally, in the military it's likely the classes you attend will relate directly to your work, whereas in the civilian world a young person often goes to college with only a vague sense of how what he or she is learning will relate to a future job.

In general, there are three ways of getting an education through the military: before you enter active duty, while you are on active duty, and when your active duty is finished. Each choice has its own advantages and disadvantages, which are discussed as follows.

Education before active duty. You might call this the "go to school now, pay later" plan. ROTC and the academies, both of which are discussed in greater detail in Chapter 5, fall into this category. The big advantage here is that you can enter college full time right after high school, as most people do. You can take classes with people your own age—maybe even people with whom you went to high school. You won't feel out of place as someone older might, and your study skills will not have had a chance to deteriorate during a prior stint of active duty.

The negative side of this arrangement is that the military dictates what you'll study. There is little choice among courses in the academies, and even in ROTC you will probably be limited to a science or engineering curriculum. This may be fine with you, but if you would rather major in English or anthropology, you may be out of luck. If you do major in a technical field, your knowledge may be out of date when you return to the civilian job market.

Another problem is that the academies and ROTC—at least for those who want four-year scholarships—are very selective, and tend to choose people who have done very well in high school. If you graduated with high honors this may be no problem, but if you didn't, you may want to choose another way of funding your education.

Finally, you will have a limited choice of schools. In some states only a handful of colleges offer ROTC programs.

Education during active duty. This is one of the more appealing arrangements: get an education and serve in the military at the same time—go now, pay now. It can be done. The armed forces go to great lengths to provide classroom space, instructors, and arrangements whereby you can accumulate college credits as you move around. The Navy, for example, dispatches college teachers on some of its ships so that sailors can attend classes while thousands of miles from the nearest shore.

The cost of getting an education while in uniform is usually low. Some programs are free, and the military pays for 90 percent of the cost of others. If you are an officer, you can go full time to graduate school or to a professional school (such as in medicine, law, or business) and still receive your paycheck.

There *are* costs in going to college while in uniform, though they may not be financial. For one thing, unless you are going to school full time a good part of your education takes place in addition to your military duties. A young person who has never worked full time may underestimate the discipline and commitment it takes to both hold down a job and go to classes and study while off duty. And people in the armed forces do not work a forty-hour week. If the task requires extra time, you have to stay there and do it. While your friends are off having a good time on the weekends, you may have to stay in and study for final exams. It's not always easy to do.

As with those who get their education before going in, the military exercises control over what and where you can study. No branch of the service is going to pay for you to go to graduate school to study a subject of no direct use to the military.

And if the military sends you to graduate school, you will have to stay in uniform longer. A good rule of thumb is that you will have to stay in two years for every academic year you spend in school.

Officers and enlisted people have a limited choice of colleges they can attend. And because of the frequent moving that military people endure, they are seldom in one place long enough to complete the requirements for a four-year degree. The military has some ways of getting around this, however. The Navy, for example, has a program in which participating colleges waive residency requirements and allow you to count credits earned at other institutions. After signing an agreement to meet all the institution's requirements, you can take as much as ten years to complete your degree, even if you leave the Navy in the meantime.

Education after active duty. The advantages of finishing your active duty before going to college center on three words: freedom of choice. Once you are finished with active duty you can take any course of study you wish at an approved institution—and virtually all established colleges are approved. You can go as a part-time, full-time, or work-study student. When you finish you can go straight into the job market with a brand new degree. It's all up to you.

The Army College Fund and the New G.I. Bill are perhaps the best ways of paying for college after your active duty is over. For just two years of active duty in the Army, you can leave with $17,000 for tuition and other costs. For four years of active duty you get $25,200, the maximum amount of money available from this program. That's not bad—how many people do you know who can save $6,300 a year?

If you postpone your education you will be better able to concentrate on doing a good job in the military. You won't have to squeeze in classes here and there, and when the weekend or leave time comes you will be ready for the good times to roll. It's called "pay now, go later."

Some people like the idea of a break between high school and college. A few years in uniform, they argue, give you a chance to see some different places, experience the world outside the classroom, and get a better sense of what you want out of life. Then, they reason, you are ready for buckling down and getting a good education. Many college instructors find that older students are more serious about their studies and therefore get more out of them.

The key phrase here is *older student*. How will you feel about being three or four years older than your classmates? After living in a barracks or an apartment in Germany, will you be content to stay in a dormitory, where beer is banned, with a bunch of eighteen-year-olds? Will it be harder to get dates with younger students? For some people being older than their classmates may be no problem at all. Others may find that they never quite fit in.

There's always the risk that once you leave school you'll never go back. It's easy to lose your study habits, and commitments you might not now predict—marriage and children, to name two big ones—can make it much more difficult for you to begin studying for tests and writing papers again.

Educational Programs for Active-Duty Personnel

The largest number of educational possibilities is offered while you are in uniform. Some programs are common to all the branches, and others are unique to one branch or another. Keep in mind that educational benefits change from time to time; check with your local recruiter if you have any questions about a particular one.

The programs described below are offered to enlistees or

officers who have been in uniform for less than four years. Each branch of the service has options for personnel who are further along their career path, but these will not be covered here.

Across-the-board programs

☐ *DANTES.* This stands for Defense Activity for Non-traditional Education Support, an agency that provides opportunities to take tests and transfer experience into credits that are recognized by civilian colleges. Among other things, DANTES offers the GED test, the ACT and SAT exams needed to apply for college, and the Graduate Record Exam (GRE) required for graduate school.

☐ *Advanced Education Program.* Approximately thirty officers annually are selected to attend a civilian university for up to twenty-four months of full-time study in pursuit of a master's degree. You get your salary and allowances, but you have to pay for the schooling and serve extra time.

☐ *Tuition Assistance Program.* For those on active duty, this program will pay up to 90 percent of the tuition for college courses if you rank between E-5 and E-9 with less than fourteen years of active service. The program will pay 75 percent of the tuition costs if you are an officer or fall elsewhere in the ranks. You must take the courses on your own time.

☐ *Servicemembers' Opportunity Colleges.* This is a group of 438 colleges and universities that have agreed to a reasonable transfer of credits among themselves for military students who move around a lot. The colleges include institutions such as Central Texas College, Pensacola Junior College, and Monterey Peninsula College, all located near military installations.

☐ *Broadened Opportunity for Officer Selection and Training (BOOST).* This program was set up to enable qualified enlisted people to receive academic preparation

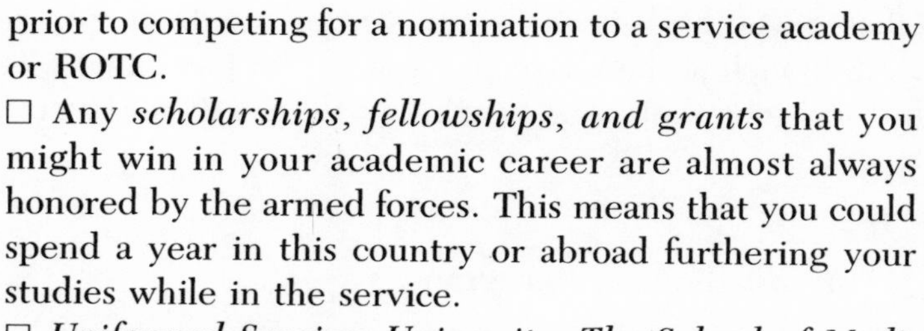

prior to competing for a nomination to a service academy or ROTC.

☐ Any *scholarships, fellowships, and grants* that you might win in your academic career are almost always honored by the armed forces. This means that you could spend a year in this country or abroad furthering your studies while in the service.

☐ *Uniformed Services University: The School of Medicine.* This is the armed forces' place to become a doctor. It costs you nothing, but you incur a seven-year obligation when you finish.

Army benefits

☐ *The Army College Fund* combined with the New G.I. Bill is one of the best educational deals in the armed forces. It works like this: you choose from a special list of jobs—many of which are combat specialties—and the Army gives you a bonus. You then contribute $100 of your monthly salary into a fund for one year, and the government contributes up to eight times that amount. The total amount is available for your education for ten years after your discharge.

Here are the numbers if you enlist for two years. You sign up for the plan, and immediately $8,000 from the Army College Fund is put into your account. Under the New G.I. Bill (see Veterans' Educational Benefits) you'll have $100 a month (approximately one-sixth of a private's wages) deducted from your pay during your first year and placed in your account. So at the end of one year you'll have added $1,200 to the original $8,000. After you serve twenty-four consecutive months the government will add $7,800, and you will walk out of active duty with a total of $17,000 available for education at an accredited college, technical school, or correspondence school.

If you enter the Army for two years with sixty hours of college credit under your belt, the Army College Fund gives you $12,000 to start off with, for a final figure of $21,000. If you enlist for three years of active duty the government contributes $12,000 initially and $9,600 later; you leave with $22,800. If you enlist for four years of active duty the government gives you $14,400 to start and $9,600 later, so you end up with $25,200. Your educational benefits are paid on a monthly basis, and they are not taxed. You must have an honorable discharge to receive any money other than your own contribution.

□ *The Health Profession Scholarship Program* will pay your way through civilian medical school and give you money to live on while you become a doctor. Once you're out you owe the Army one year for every year you were in this program.

□ *The Fully Funded Legal Education Program* is just what the name says; it will pay your way through law school. You have to be an officer with at least two but no more than six years of active duty at the time the schooling begins, and when you get your degree you owe the service one year for each year spent hitting the books.

Navy benefits

□ *Enlisted Commissioning Program.* This program enables enlisted people to make the jump to the officer ranks. They attend a college in a Naval ROTC program, receiving full pay and allowances, but they have to pay for the schooling. They go to school year-round for no longer than thirty-six months, and on receiving their commission they are obligated to serve for four years.

□ *Enlisted Education Advancement Program.* Enlisted people can complete an associate degree through this program, which gives them time off with pay to go to an approved college. The individual pays the tuition and other costs and must serve an extra six years.

□ *Navy Campus High School Studies Program.* Basics

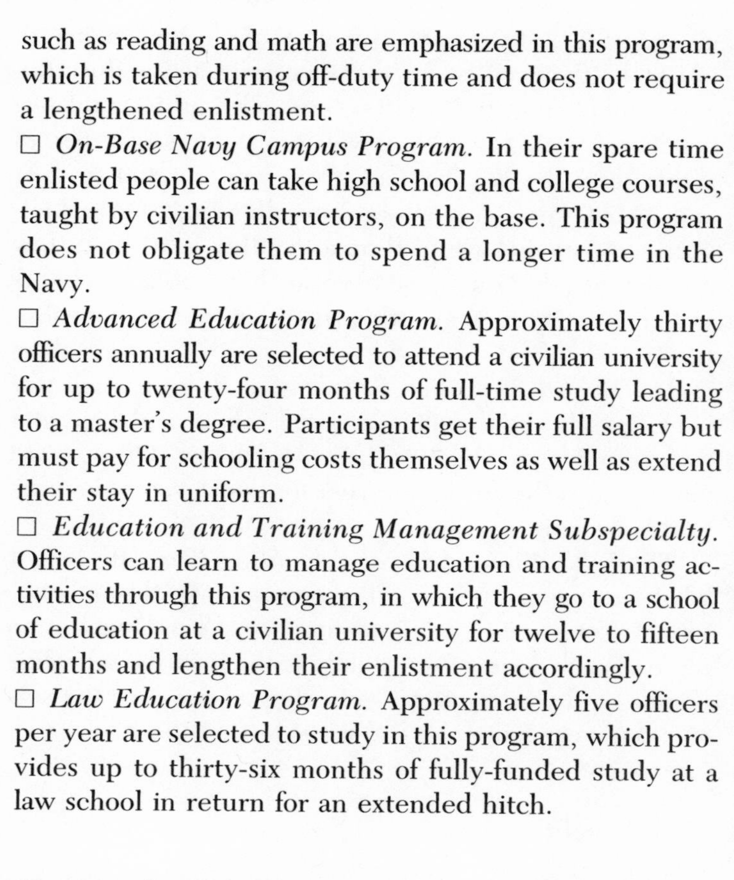

such as reading and math are emphasized in this program, which is taken during off-duty time and does not require a lengthened enlistment.

☐ *On-Base Navy Campus Program.* In their spare time enlisted people can take high school and college courses, taught by civilian instructors, on the base. This program does not obligate them to spend a longer time in the Navy.

☐ *Advanced Education Program.* Approximately thirty officers annually are selected to attend a civilian university for up to twenty-four months of full-time study leading to a master's degree. Participants get their full salary but must pay for schooling costs themselves as well as extend their stay in uniform.

☐ *Education and Training Management Subspecialty.* Officers can learn to manage education and training activities through this program, in which they go to a school of education at a civilian university for twelve to fifteen months and lengthen their enlistment accordingly.

☐ *Law Education Program.* Approximately five officers per year are selected to study in this program, which provides up to thirty-six months of fully-funded study at a law school in return for an extended hitch.

Air Force benefits

☐ *The Airman Education and Commissioning Program* enables enlisted personnel to join the officer corps. You have to have earned forty-five semester hours in college with a 2.0 (*C*) average, and you must be less than thirty-five years old when commissioned. You are sent to college to get a bachelor of science degree in a scientific or technical discipline, and then you go to OTS. You must commit yourself to six years in the Air Force when you enter the program and four years upon commissioning as an officer.

☐ *Bootstrap Temporary Duty* assigns you to full-time study at a civilian college for up to one year while you complete a bachelor of science degree. The Air Force pays your regular wages but you pay for the schooling, and once you finish you owe the Air Force three months for every one month you were in school. Following this you may apply for OTS.

☐ *The Community College of the Air Force* is the only federal agency that has the power to grant degrees. You receive academic credit for basic training and job training and then add courses taken off-duty at civilian colleges, for which the Air Force pays 90 percent of the tuition. You eventually receive an associate's degree, and no obligation is incurred.

☐ *Air Force Institute of Technology.* This program offers officers education on the bachelor, master's, and doctoral levels in various schools of engineering and systems and logistics.

☐ *Health Education Programs.* These include degree, post-graduate, and certification programs for career Air Force Medical Service officers; medical, dental, and veterinary education programs for other Air Force officers; and the Health Professions Scholarship Program.

☐ *Senior Commander-Sponsored Education Program.* Each year approximately fifty officers with strong academic and military credentials are allowed to take courses leading to master's or doctoral degrees in science, engineering, and technical management.

☐ *Education with Industry Program.* Unlike the other programs for officers, this one does not lead to a degree. Selected officers are assigned to industries, governmental agencies, and businesses for managerial training and experience in such fields as computer resources, cost analysis, audiovisual materials, retail, transportation, and nuclear engineering.

Marine Corps benefits

☐ *Enlisted Commissioning Education Program.* Selected Marines aged twenty to twenty-six can attend college full

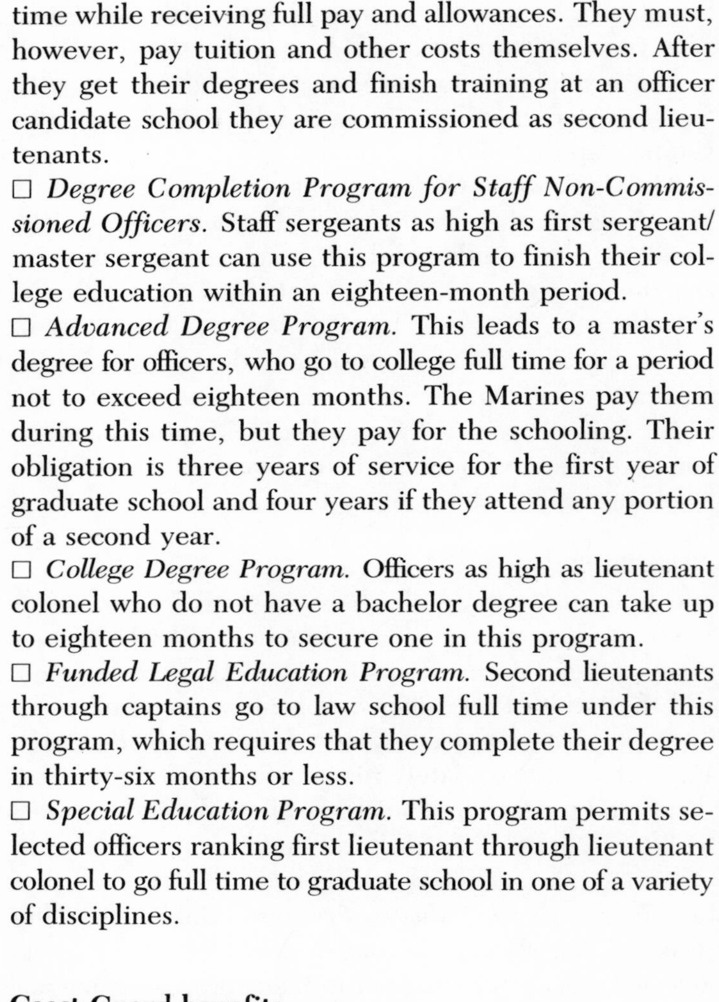

time while receiving full pay and allowances. They must, however, pay tuition and other costs themselves. After they get their degrees and finish training at an officer candidate school they are commissioned as second lieutenants.

□ *Degree Completion Program for Staff Non-Commissioned Officers.* Staff sergeants as high as first sergeant/master sergeant can use this program to finish their college education within an eighteen-month period.

□ *Advanced Degree Program.* This leads to a master's degree for officers, who go to college full time for a period not to exceed eighteen months. The Marines pay them during this time, but they pay for the schooling. Their obligation is three years of service for the first year of graduate school and four years if they attend any portion of a second year.

□ *College Degree Program.* Officers as high as lieutenant colonel who do not have a bachelor degree can take up to eighteen months to secure one in this program.

□ *Funded Legal Education Program.* Second lieutenants through captains go to law school full time under this program, which requires that they complete their degree in thirty-six months or less.

□ *Special Education Program.* This program permits selected officers ranking first lieutenant through lieutenant colonel to go full time to graduate school in one of a variety of disciplines.

Coast Guard benefits

□ *Correspondence Courses.* The Coast Guard offers courses by mail in a variety of subjects.

□ *Advanced Electronics Associate Degree Program.* Enlisted people of E-6 rank or above are paid to go to community college for two years to receive an associate's degree in electronics. The obligation is two years of service for every year in school.

□ *Officer Post Graduate Program.* The Coast Guard

gives officers the opportunity to go to graduate school in fields that include civil engineering, physics, and law. The obligation incurred is two years for every year spent in school.

POST-SERVICE BENEFITS

The American public and Congress have traditionally felt that military personnel deserve more than just a discharge when their active duty is finished. This gratitude takes the form of veterans' benefits, which are administered by the Veterans Administration (V.A.).

Not surprisingly, those who have made the military their career get the best deal. If you serve in the armed forces for twenty years, you can retire and receive a pension that equals half of your final salary. This means that some people can retire at age thirty-eight or thirty-nine, with plenty of time to move into another career, and live the rest of their working lives on a paycheck and a half. Those who stay in for thirty years can retire on 75 percent of their salary.

Veterans who have been wounded or injured while on active duty also receive favorable treatment. Besides getting free medical care, they are eligible for pensions covering various degrees of disability. For those killed in action or while on active duty, benefits go to their spouses and children.

But you don't have to be a twenty-year member or a war hero to reap veterans' benefits. Many advantages are yours even if the only shots you ever fired were on the rifle range. Chief among these are educational and employment benefits.

Veterans' Educational Benefits

After World War II the Serviceman's Adjustment Act, commonly known as the G.I. Bill, enabled thousands of veterans to go to college.

The current version of this legislation, called the New G.I. Bill, works like this: unless you object, the military will deduct $100 from your monthly wages during your first year in uniform. If you serve twenty-four consecutive months, the government will add $7,800 to your contribution. Provided you are honorably discharged, you wind up with $9,000 for college costs.

This money can be applied to your education for up to ten years after your discharge. The V.A. will pay you a portion of the amount every month you are in school. If you go to college for four years and attend classes nine months a year, you will receive a check for $250 each of those nine months. If you decide to go to school part time, your monthly amount will be reduced, but the benefits will be paid over a longer period as your years of schooling are extended.

This assumes you contribute only $1,200 toward your education. You can always save more of your salary, as well as any bonuses you might receive, on your own.

Veterans' Employment Benefits

Whether you go to college or not you can take advantage of "veteran's preference," a practice whereby former service people are given special help in securing government jobs.

This can work a couple of ways. Applicants for many governmental jobs are required to take a civil service exam, the results of which are used to rank the people who want the position. Veterans are often given a number of points to add to their score on the test; this puts them higher on the list and makes it more likely that they will get the job. In other places simply being a veteran and qualified for the job is enough to land it.

Some employers in the private sector, particularly former servicemen, have their own versions of veteran's preference. If two job applicants are equal in every way and one of them is a veteran, he or she will get the post.

Other Veterans' Benefits

The V.A. can help you obtain housing. Provided you and the house or mobile home you wish to purchase meet certain standards, the V.A. will guarantee up to 60 percent of a loan to a maximum of $27,500. This loan guarantee can sometimes be just the help needed to buy a home.

Under certain circumstances veterans who haven't been wounded in action are eligible for free medical care. The V.A. operates hospitals in most parts of the country. You will be admitted if space is available and you cannot pay for treatment elsewhere.

Finally, veterans can be buried in national cemeteries or receive an allowance for burial expenses elsewhere. The V.A. will also provide an American flag for the funeral and a headstone or grave marker.

16. A Final Choice

Photo by author

IN TALKING ABOUT ENLISTMENTS, officer training and everything else, this book has so far limited its concern to people who *want* to enter the armed forces. All young men who reach the age of eighteen, however, must consider the military for one reason—the draft.

Registration and the Draft

Talk about a revival of the draft surfaces periodically among military leaders, governmental figures, and the general public. With the registration of eighteen-year-old men some people saw the revival of the draft as imminent, but there has been little official movement in this direction.

As long as the military has no trouble finding enough recruits, the situation will probably stay as it is. Census figures, however, show that the number of military-age males is declining and will continue to decline until the mid-1990s. This drop in available recruits, plus the changes in the national economy, will greatly affect whether or not the United States returns to a draft.

If the economy improves and more civilian jobs become available, people will be less likely to join the service. But if federal deficits continue unchecked, another recession could make young people line up at recruiting offices once more.

One compelling argument for a draft has nothing to do with census figures or the state of the economy. This point of view contends that all young people should serve their country in some fashion, military or otherwise. Now, as in the past, people with less money have been far more likely to find themselves in uniform than those with more. Some argue that a program of national service would be good for both the country and the individual. This line of thought has support from political conservatives as well as liberals.

In all this discussion and uncertainty one thing is sure: the government expects young men to register for the draft at their local post offices. Once you fill out the form it is sent to the headquarters of the Selective Service. There are no deferments at this stage of the game; whether you are a pacifist or totally gung-ho, you are required to register.

Unless you are female. While feminists declare that a male-only draft is unfair, the Supreme Court ruled that it is at least constitutional. There seems to be no groundswell of opinion in favor of drafting females; those women who are interested in a military career are more concerned with seeing the ban on combat lifted.

Some people oppose registration and advise young men not to participate. These individuals and organizations reason that if enough males refuse to register it will be impossible for the government to prosecute them all. So far the government hasn't tried. Those young men who have not registered have generally gone unprosecuted, unless they have taken public stances or somehow provoked the government into action. But every resister takes the risk of being convicted and subjected to the maximum penalties—five years in prison and a $10,000 fine.

If you decide to resist the draft or make plans to apply for conscientious objector status, peace groups advise that you do the following:

☐ *Sort out your beliefs* concerning war and military service. If you are called in to explain why you are seeking conscientious objector status, you will be closely questioned about what you believe and why.

☐ *Talk to a draft counselor,* who can help you arrive at a decision. He or she will be familiar with military procedures and regulations and may be a veteran or a political activist. You can locate a draft counselor by looking in the phone book, asking a minister, or contacting one of the groups listed in Chapter 12, "How to Get Out."

A draft counselor can give you the pros and cons of various plans of action. But one word of caution: much like an overzealous recruiter who stretches the truth to sign you up, some draft counselors may push you to take a more radical stance than you might otherwise.

☐ *Register with a church or political group* as a conscientious objector. This puts you on record as having these beliefs and may help you prove your case later on.

A Final Word

Military service offers a wide range of opportunities for young people. Some of the benefits are widely advertised, such as the attainment of job skills and money for education. Others aren't talked about so much, such as the chance to make fundamental changes in your life.

In considering whether or not to join, you need to examine the opportunities and yourself and find out just what you want from the military. Is it a chance to get away from home? Is it an education? Or is it a change from a boring life? Whatever your reasons, you need to look around and see if there is any other place you can satisfy your needs.

If you want to get away from home, there are jobs that involve lots of travel. If you want a good education, there are plenty of scholarships and schools where you can work your way through. These places don't have millions of dollars to advertise themselves every day, but you can find them.

Talk to people. Many young people never seek advice from anyone besides their contemporaries—those who have had no more experience than themselves. If you want to go to college, spend some time talking with your high school counselor. If he or she cannot help, go to the financial aid office of a local college. Or do some reading on the matter.

If you are looking for job skills, talk to people who are doing the work you want to do; find out how they got their jobs. Talk to representatives of labor unions in your area and people in unemployment offices. You might locate an apprenticeship program that does not ask you to go to boot camp and to sign up for eight years.

All of this takes time and effort, and it's never easy. Some people won't take time to talk with you; others may be friendly but of little help. You may live in an area of few resources. Just the same, it's a good idea to explore all the options before talking to a recruiter. You may wind up in uniform anyway, but at least you will know you made the best choice.

When you do at last go to a recruiter, don't be in any rush to sign up. Joining the military will profoundly affect your life for the next eight years. This is something you shouldn't do on the spur of the moment.

As you compare the various branches of the service and the programs they offer, seek out veterans of each and see what they have to say. Veterans will usually be more frank with you than any recruiter. Ask them what the advantages and disadvantages were for them, and get them to tell you what they would do if they were to enlist again.

Once you narrow your list down to one or two branches and begin talking about jobs, press the recruiter for more information. Don't just accept the job title and a vague description. In the Army, for example, one title among the building trades is interior electrician—something that sounds very useful in civilian life. You do inside wiring in this job, but you may also be asked to lay and clear mine fields, prime and place explosives, and perform other combat-related tasks. This is not apparent from the name of the job or the description.

Find out what it takes to get the job you are interested in. Ask about the requirements and the tests. You may find that learning the skill you want requires qualifications that you don't have. If you get this information before you go in, *you* can decide what else to try. If you are already in uniform when this is discovered, *they* decide what you will do.

In short, before you sign any papers you should have all the information you can get about where you are going, what you are going to do, and how you will do it. Try to minimize the surprises.

This book has pointed out the negative aspects of a military commitment as well as the positive ones, but you should keep in mind that the civilian world isn't perfect either. Racism and sexism exist in industry and offices as surely as they do in boot camp, and all too often people who aren't in uniform are promised one thing and delivered another. The big difference is that with almost any civilian situation if things don't go your way you can quit. It is possible to get out of the military before your hitch is over, but it is extremely difficult to do so.

Finally, military service is just that—service to the country. It cannot be compared to a civilian job. There are benefits and job descriptions and other similar features, but few civilian employers will ask you to do the things that the military will.

If you join the armed forces it is possible that you will be sent into combat. You may find yourself fighting in a war you do not support, against people you do not consider enemies. Or you may proudly take part in defending our country and our way of life, or lending support to people who strive for the same things we cherish. Either way, you may end up seriously hurt or even killed. Many professional military people refer to their mission as a calling. It can certainly be that—but it's not for everyone.

If you want to go into the armed forces, you owe it to yourself to make sure it's the best thing for you.

Appendix: Colleges and Universities Offering ROTC

The following lists indicate where various ROTC programs are offered. The main entries for each state are host institutions; these are where classes and drills are held, and where the ROTC professors have their offices. The subentries are nearby colleges with a cross-enrollment agreement, whereby their students can enroll in the ROTC courses at the host institution.

In picking a college or an ROTC program, keep these things in mind. First, you should attend a host institution if you can. If this is not possible, pick a cross-enrollment school that is close to the host college. You may have to go to ROTC classes and drills as often as three times a week; a long commute could eat up time and cause scheduling problems with your other classes.

Second, from year to year schools halt cross-enrollment programs, host institutions change, and new colleges are added and dropped. Before enrolling anywhere you should check with a professor of military science at a host college or write the headquarters of each branch to make sure the program you have in mind is still there. Here are the addresses:

□ *Army.* Army ROTC, Fort Monroe, Virginia 23651

□ *Navy and the Marine Corps.* Navy Recruiting Command (Code 314), 4015 Wilson Boulevard, Arlington, Virginia 22203–1991

□ *Air Force.* AFROTC Recruiting Division, Maxwell Air Force Base, Alabama 36112

Army

ALABAMA

Alabama A & M University
- Athens State College
- John C. Calhoun State Community College
- Oakwood College
- University of Alabama—Huntsville

Auburn University
- Alabama State University
- Auburn University at Montgomery
- Huntington College
- Troy State University

University of North Alabama

Jacksonville State University
- Talladega College

Marion Military Institute

Tuskegee University

University of Alabama
- Stillman College

University of Alabama at Birmingham
- Birmingham Southern College
- Miles College
- Samford University
- University of Montevallo

University of South Alabama
- Mobile College
- Spring Hill College

ALASKA

University of Alaska—Fairbanks

ARIZONA

Arizona State University
- Glendale Community College
- Mesa Community College
- Phoenix College
- Scottsdale Community College

Northern Arizona University
- Emery-Riddle Aeronautical University

University of Arizona
- Pima Community College

ARKANSAS

Arkansas State University

Arkansas Technical University

College of the Ozarks

Henderson State University

Ouachita Baptist University

Southern Arkansas University

University of Arkansas
- Northeastern State University (Oklahoma)

University of Arkansas—Little Rock
- Philander Smith College
- University of Arkansas—Medical Science Campus

University of Arkansas at Pine Bluff
- University of Arkansas at Monticello

University of Central Arkansas
- Central Baptist College
- Hendrix College

CALIFORNIA

California Polytechnic State University
- Cuesta College

California State University—Fresno

California State University—Long Beach
- California State University—Dominguez Hills
- Cerritos College
- Cypress College
- El Camino College
- Fullerton College
- Glendale Community College
- Golden West College
- Orange Coast College
- Santa Ana College
- University of California—Irvine

Claremont Colleges
- Art Center College of Design
- Azusa Pacific College
- California Baptist College
- California Institute of Technology
- California State Polytechnic University
- California State University—Fullerton
- California State University—Los Angeles
- California State University—San Bernardino
- Cerritos College
- Chaffey College
- Citrus College
- College of the Desert
- Crafton Hills College
- Cypress College
- Fullerton College
- Loma Linda University
- Mount San Antonio College
- Pasadena City College
- Point Loma Nazarene College
- Rio Honda College
- Riverside City College
- Saddleback Community College
- San Bernardino Valley College
- Santa Ana College
- University of California—Irvine
- University of California—Riverside
- University of La Verne
- University of Redlands
- Victory Valley College
- Whittier College

San Diego State College
- Cuyamaca College
- Grossmont College
- National University
- Palomar College
- Point Loma College
- San Diego City College
- San Diego Mesa College
- San Diego Miramar College
- Southwestern College
- University of California—San Diego
- University of San Diego

San Jose State University
- Cabrillo College
- College of San Mateo
- De Anza College
- Evergreen Valley College
- Foothill College
- Gavilan Community College
- Hartnell College
- Mission College
- Monterey Peninsula College
- Ohlone College
- San Jose City College
- West Valley College

University of California—Berkeley
- California State University—Hayward
- Chabot College
- College of Alameda
- Contra Costa College
- Diablo Valley College
- Golden Gate University
- John F. Kennedy University
- Laney College
- Los Mecanos College
- Merritt College
- Mills College
- Napa College

San Francisco Community College
San Francisco State University
San Mateo County Community College
Solano Community College
St. Mary's College of California

University of California—Davis
American River College
California State University—Chico
California State University—Sacramento
Sacramento City College

University of California—Los Angeles
Antelope Valley College
Azusa Pacific College
California Institute of the Arts
California State University—Dominguez Hills
California State University—Los Angeles
California State University—Northridge
Cerritos College
College of the Canyons
Cypress College
El Camino College
Fullerton College
Glendale Community College
Long Beach City College
Los Angeles City College
Los Angeles Mission College
Los Angeles Pierce College
Los Angeles Southwest College
Los Angeles Trade Technical College
Los Angeles Valley College
Northrop University
Occidental College
Orange Coast College
Pepperdine University
Saddleback Community College
Santa Ana College
Santa Monica College
University of California—Irvine
University of Southern California
University of West Los Angeles
Valley College of Medicine Center
Ventura College
West Los Angeles College

University of California—Santa Barbara
California Lutheran College
Westmont College

University of San Francisco
California State University of Hayward
City College of San Francisco
College of Marin
College of San Mateo
Dominion College of San Rafael
San Francisco State University
Santa Rosa Junior College
Skyline College
St. Marus College of San Francisco

University of Santa Clara
De Anza College
Foothill College
Stanford University

University of Southern California
California State University—Los Angeles
Don Bosco Technical Institute
Mount St. Mary's College
Occidental College
Rio Hondo College
West Los Angeles College
Woodbury University

COLORADO

Colorado School of Mines

Colordo State University
Northeastern Junior College
University of Northern Colorado

Metropolitan State College
Front Range Community College
Loretta Heights College
Mesa College
Regis College
University of Colorado Health Sciences Center
University of Colorado—Denver
University of Denver

University of Colorado at Boulder
Academy for Nursing Science

University of Southern Colorado
Adams State College

CONNECTICUT

University of Connecticut
Annhurst College
Central Connecticut State College
Eastern Connecticut State College
Fairfield University
Greater Hartford Community College
Housatonic Regional Community College
Mattatuck Commmunity College
Middlesex Community College
Northwestern Connecticut Community College
Sacred Heart University
St. Joseph College
South Central Community College
Southern Connecticut State College
Trinity College
University of Bridgeport
University of Connecticut—Hartford Branch
University of Hartford
University of New Haven
Western Connecticut State College
Yale University

DELAWARE

University of Delaware
Delaware State College
Lincoln University (Pennsylvania)
Salisbury State College (Maryland)
Wilmington College

DISTRICT OF COLUMBIA

Georgetown University
American University
Catholic University of America
Columbia Union College (Maryland)
George Mason University (Virginia)
George Washington University
Marymount College of Virginia (Virginia)
Mount Vernon College

District of Columbia *cont.*
- Northern Virginia Community College (Virginia)
- Trinity College
- University of Maryland—College Park Campus
- University of the District of Columbia

Howard University
- Anne Arundel Community College (Maryland)
- Baltimore Hebrew College (Maryland)
- Bowie State College (Maryland)
- Capitol Institute of Technology (Maryland)
- Coppin State College (Maryland)
- George Washington University
- Prince George's Community College (Maryland)
- University of the District of Columbia
- University of Maryland (Maryland)
- Virginia Union University (Virginia)

FLORIDA

Florida Southern College
- Hillsborough Community College
- Polk Community College
- St. Leo College
- Southeastern College—Assemblies of God
- Webber College

Florida State University
- Chipola Junior College
- Gulf Coast Community College
- Pensacola Junior College
- Tallahassee Community College
- University of West Florida

Stetson University
- Central Florida Community College
- Seminole Community College
- Valencia Community College

University of Florida
- Edward Waters College
- Flagler College
- Florida Junior College—Jacksonville
- Jacksonville University
- St. John's River Community College
- Santa Fe Community College
- University of North Florida

University of Miami
- Barry College
- Biscayne College
- Broward Community College
- Florida Atlantic University
- Florida International University
- Florida Memorial College
- Miami-Dade Community College
- Nova University

University of South Florida
- Eckerd College
- Hillsborough Community College
- St. Leo College
- St. Petersburg Junior College
- University of South Florida, St. Petersburg

University of Tampa
- Hillsborough Community College
- St. Petersburg Junior College

GEORGIA

Augusta College
- Georgia College
- Medical College of Georgia
- Paine College
- University of South Carolina—Aiken

Columbus College
- Andrew College
- Chatahoochie Valley Community College
- Emmanuel College
- La Grange College

Fort Valley State College
- Albany State College

Georgia Institute of Technology
- Berry College
- Clark College
- Emory University
- Floyd Junior College
- Kennesaw College
- Morehouse College
- Morris Brown College
- Shorter College
- Spelman College

Georgia Military College
- Georgia College

Georgia Southern College
- Armstrong State College
- Emanuel County Junior College
- Savannah State College

Georgia State University
- Atlanta Junior College
- Atlanta University
- Clayton Junior College
- DeKalb Community College
- Gordon Community College
- Kennesaw College
- Mercer University—Atlanta
- West Georgia College

Mercer University
- Georgia Southwestern College
- Middle Georgia College

North Georgia College
- Young Harris College

University of Georgia
- DeKalb Community College
- Emmanuel College
- Floyd Junior College
- Gainesville Junior College
- Paine College
- Piedmont College
- Savannah State College
- Shorter College
- Truett McConnell College

HAWAII

University of Hawaii
- Brigham Young University of Hawaii
- Chaminade University of Hawaii
- Hawaii Loa College
- Hawaii Pacific College
- Honolulu Community College
- Kapiolani Community College
- Leeward Community College
- West Oahu College
- Windward Community College

ILLINOIS

Chicago State University
- Governor's State College
- Indiana University—Northwest

Eastern Illinois University

Illinois State University
- Illinois Wesleyan University

Knox College
- Bradley University
- Carl Sandburg College
- Illinois Central College
- Monmouth College

Loyola University of Chicago

Barat College
Columbia College
De Paul University
Devry Institute of Technology
Illinois Institute of Technology
Lake Forest College
Northeastern Illinois University
Northwestern University
University of Chicago
Northern Illinois University
Kishwaukee College
Southern Illinois University—Carbondale
Southeast Missouri State University (Missouri)
Western Illinois University
Black Hawk College
Spoon River Community College
Wheaton College
Aurora University
College of Du Page
Devry Institute of Technology
Elgin Community College
Elmhurst College
Governor's State University
Illinois Benedictine College
Joliet Junior College
Judson College
Lewis University
Moraine Valley Community College
North Central College
Olivet Nazarene College
Roosevelt University
Triton College
William Rainey Harper College
University of Illinois
Parkland College
University of Illinois—Chicago
City College of Chicago—Kennedy-King Branch
City College of Chicago—Loop Branch
City College of Chicago—Malcolm X Branch
City College of Chicago—Truman Branch
City College of Chicago—Wright Branch
College of Du Page
College of Lake County
Illinois Institute of Technology
Moraine Valley Community College
Rush University
St. Xavier College
University of Illinois—Medical Center

INDIANA
Ball State University
Indiana University—Fort Wayne
Marion College
Indiana University
Indiana University—Southeast
Indiana University, Purdue University—Indianapolis
Butler University
DePauw University
Franklin College of Indiana
Indiana University at Kokomo
Marian College
University of Indianapolis
Purdue University
Rose-Hulman Institute of Technology
DePauw University
Indiana State University
St. Mary-of-the-Woods
Vincennes University
University of Notre Dame
Bethel College
Holy Cross Junior College
Indiana University—South Bend
Purdue University—Calumet
St. Mary's College

IOWA
Iowa State University
Des Moines Area Community College
Drake University
Grand View College
University of Northern Iowa
University of Iowa
Coe College
Iowa Western Community College
Kirkwood Community College

KANSAS
Pittsburg State University
Kansas State University
University of Kansas
Baker University
Emporia State University
Washburn University
Wichita State University
Fort Hays State University
Saint Mary of the Plains College

KENTUCKY
Eastern Kentucky University
Cumberland College
Union College
Morehead State University
Murray State University
University of Kentucky
Center College of Kentucky
Georgetown College
Kentucky State University
Transylvania University
University of Louisville
Western Kentucky University

LOUISIANA
Louisiana State University, A & M College
Louisiana State University at Alexandria
Loyola University
McNeese State University
Lamar University (Texas)
University of Southwestern Louisiana
Nicholls State University
Northeast Louisiana University
Grambling State University
Northeast State University of Louisiana
Centenary College
Louisiana College
Louisiana State University—Shreveport
Southeastern Louisiana University
Tulane University
Dillard University
Louisiana State University—Medical Center
Southern University in New Orleans
University of New Orleans
Xavier University

MAINE
University of Southern Maine
St. Joseph's College
Westbeck College
University of Maine
Husson College
Nasson College
Thomas College

MARYLAND

The Johns Hopkins University
- Anne Arundel Community College
- Catonsville Community College
- Community College of Baltimore
- Coppin State College
- Goucher College
- Howard Community College
- Maryland Institute College of Art
- Prince George's Community College
- Towson State University
- University of Baltimore
- University of Maryland—Baltimore County
- University of Maryland—College Park
- University of Maryland—Professional

Loyola College
- Catonsville Community College
- College of Notre Dame—Maryland
- Community College of Baltimore
- Essex Community College
- Harford Community College
- Towson State University
- University of Baltimore
- University of Maryland
- University of Maryland—Baltimore County

Morgan State University
- Coppin State College

Western Maryland College
- Catonsville Community College
- Essex Community College
- Frederick Community College
- Hood College
- Shephard College (West Virginia)
- Towson State University
- University of Maryland—Baltimore County
- University of Maryland—College Park

MASSACHUSETTS

Boston University
- Anna Maria College
- Babson College
- Bentley College
- Boston College
- Brandeis University
- Bridgewater State College
- Cape Cod Community College
- Framingham State College
- Massasoit Community College
- Middlesex Community College
- Northeastern University
- Southeastern University
- Stonehill College
- University of Massachusetts—Boston
- Westfield State College
- Wheaton College

Massachusetts Institute of Technology
- Harvard University
- Tufts University
- Wellesley College

Northeastern University
- Bentley College
- Berklee School of Music
- Boston College
- Bradford College
- Brandeis University
- Bristol Community College
- Bunker Hill Community College
- Curry College
- Eastern Nazarene College
- Emmanuel College
- Framingham State College
- Gordon College
- Massachusetts Bay Community College
- Massachusetts College of Art
- Massachusetts College of Pharmacy
- Merrimack College
- Middlesex Community College
- New England School of Law
- Northern Essex Community College
- North Shore Community College
- Quincy Junior College
- Roxbury Community College
- Salem State College
- Simmons College
- Suffolk University
- University of Lowell
- University of Massachusetts—Boston
- Wentworth Institute of Technology

University of Massachusetts
- American International College
- Berkshire Community College
- College of Our Lady of Elms
- Holyoke Community College
- Springfield College
- Springfield Technical Community College
- Western New England College
- Westfield State College

Worcester Polytechnic Institute
- Anna Maria College
- Assumption College
- Central New England College
- Clark University
- College of the Holy Cross
- Fitchburg State College
- Framingham State College
- Hellenic College
- Merrimack College
- Middlesex Community College
- Mount Wachusett Community College
- Nichols College
- Northern Essex Community College
- Quinsigamond Community College
- University of Lowell
- Worcester State College

MICHIGAN

Central Michigan University
- Alma College
- Ferris State College
- Northwood Institute
- Saginaw Valley State College

Eastern Michigan University

Michigan State University
- Lansing Community College

Michigan Technical University

Northern Michigan University

University of Detroit
- Detroit College of Business Administration
- Henry Ford Community College
- Highland Park Community College
- Lawrence Institute of Technology

Macomb Community College—Center Campus
Madonna College
Marygrove College
Mercy College of Detroit
Oakland Community College
Oakland University
Walsh College
Wayne County Community College
Wayne State University
University of Michigan
Adrian College
University of Michigan—Dearborn
Western Michigan University
Albion College
Calvin College
Davenport College
Grand Rapids Baptist College
Grand Valley State College
Hope College
Kalamazoo College
Kalamazoo Valley Community College
Kellogg Community College
Nazareth College
Olivet College

MINNESOTA
Bemidji State University
Mankato State University
Bethany Lutheran College
Gustavus Adolphus College
St. Johns University
College of St. Benedict
Crosier Seminary
St. Cloud State University
University of Minnesota
Anoka Ramsey Community College
Augsburg College
Bethel College
College of St. Thomas
Concordia College—St. Paul
Gustavus Adolphus Collge
Hamline University
Inver Hills Community College
Lakewood Community College
Macalester College
Metropolitan Community College
Normandale Community College
North Hennepin Community College
Northwestern College
University of Wisconsin—River Falls (Wisconsin)
University of Wisconsin—Stout (Wisconsin)

MISSISSIPPI
Alcorn State University
Delta State University
Mississippi Valley State University
Jackson State University
Millsap College
Mississippi College
Tougaloo College
Mississippi State University
Meridian Junior College
University of Mississippi
University of Southern Mississippi
William Carey College

MISSOURI
Central Missouri State University
Kemper Military School and College
Central Methodist College
Lincoln University
Missouri Western State College
Benedictine College (Kansas)
Park College
Rookhurst College
University of Missouri—Kansas
Northeast Missouri State University
Northwest Missouri State University
Southwest Missouri State University
Baptist Bible College
Central Bible College
Drury College
Evangel College
Missouri Southern State College
Southwest Baptist University
University of Missouri—Columbia
Columbia College
Stephens College
University of Missouri—Rolla
Washington University
Fontbonne College
Harris-Stowe State College
Lindenwood Colleges
Maryville College
Missouri Baptist College
St. Louis College of Pharmacy
St. Louis Community College—Florissant Valley
St. Louis Community College—Forest Park
St. Louis Community College—Meramec
St. Louis University
St. Louis University Parks
Southern Illinois University—Edwardsville (Illinois)
University of Missouri—St. Louis
Webster University
Wentworth Military Academy and Junior College
Westminster College
William Woods College

MONTANA
Montana State University
Eastern Montana College
Montana College of Mineral Science and Technology
University of Montana

NEBRASKA
Creighton University
Bellevue College
College of St. Mary's
Iowa Western Community College
Metropolitan Technical Community College
Peru State College
University of Nebraska—Medical Center
University of Nebraska—Omaha
Kearney State College
University of Nebraska
Concordia Teacher's College
Doane College
Nebraska Wesleyan University

NEVADA
Univesity of Nevada—Las Vegas
Clark County Community College
University of Nevada—Reno

NEW HAMPSHIRE
University of New Hampshire
Daniel Webster College
Dartmouth College

New Hampshire *cont.*
- Keene State College
- Nathaniel Hawthorne College
- New England College
- New Hampshire College
- Plymouth State College
- Rivier College
- St. Anselm's College

NEW JERSEY

Princeton University
- Rider College
- Burlington County College
- Mercer County Community College
- Stockton State College
- Trenton State College

St. Peter's College
- Dominican College of Blauvelt (New York)
- Glassboro State College
- Hudson County Community College
- Jersey City State College
- Kean College of New Jersey
- Rutgers University—Newark
- Stevens Institute of Technology
- William Paterson College

Seton Hall University
- Bergen Community College
- Bloomfield College
- Caldwell College
- CUNY—John Jay College of Criminal Justice (New York)
- County College of Morris
- Drew University
- Essex County College
- Fairleigh Dickinson University—Rutherford
- Fairleigh Dickinson University—Teaneck-Hackensack
- Fairleigh Dickinson University —Madison
- Greensboro College
- Kean College of New Jersey
- Middlesex County College
- Montclair State College
- New Jersey Institute of Technology
- New York University
- Passaic County Community College
- Ramapo College of New Jersey
- Rutgers University—Newark
- Somerset County College
- Union College
- Union County Technical Institute
- Upsala College
- William Paterson College

NEW MEXICO

Eastern New Mexico University
- New Mexico Highlands University

New Mexico Military Institute

New Mexico State University
- College of Santa Fe
- New Mexico Institute of Mining and Technology
- University of New Mexico

NEW YORK

Canisius College
- Bryant Stratton Business Institute
- Daemen College
- D'Youville College
- Erie Community College
- Medaille College
- SUNY—Buffalo
- SUNY—Buffalo Main Campus
- Villa Maria College of Buffalo

Clarkson University
- Hudson Valley Community College
- SUNY—Plattsburgh
- SUNY—Potsdam

Cornell University
- Broome Community College
- Corning Community College
- Elmira College
- Hartwick College
- Ithaca College
- SUNY Agricultural and Technical College—Delhi
- SUNY—Binghampton
- SUNY—Cortland
- SUNY—Oneonta
- Tompkins-Cortland Community College

Fordham University
- Academy of Aeronautics
- Baruch College
- CUNY—Bronx Community College
- CUNY—City College
- CUNY—Hunter College
- CUNY—John Jay College of Criminal Justice
- CUNY—Lehman College
- CUNY—Queens College
- College of Mount St. Vincent
- College of New Rochelle
- Columbia University
- Columbia University Teachers College
- Dutchess Community College
- Fashion Institute
- Iona College
- Manhattan College
- Marist College
- Marymount College
- Mercy University
- Mount St. Mary's College
- New School for Social Research
- New York Institute of Technology
- New York University
- Pace University
- Rockland Community College
- School of Visual Arts
- SUNY—New Paltz
- SUNY—Purchase
- Westchester Community College

Hofstra University
- Adelphi University
- Dowling College
- CUNY—John Jay College of Criminal Justice
- CUNY—Queensborough Community College
- Long Island University—C. W. Post Center
- Long Island University—Southampton Center
- Molloy College
- Nassau Community College
- New York Institute of Technology—Main Campus
- Sh'or Yeshiva Rabbinical College
- SUNY—Old Westbury
- SUNY—Stony Brook Main Campus
- SUNY Agricultural and Technical College—Farmingdale
- Suffolk County Community College
- Westchester Community College

Niagara University
- Hartwick College

- Niagara County Community College
- Onondaga Community College

Polytechnic Institute of New York
- Brooklyn Law School
- Cathedral College of the Immaculate Conception
- CUNY—Bernard M. Baruch College
- CUNY—Borough of Manhattan Community College
- CUNY—Bronx Community College
- CUNY—Brooklyn College
- CUNY—City College
- CUNY—Staten Island College
- CUNY—Hunter College
- CUNY—John Jay College of Criminal Justice
- CUNY—Kingsborough Community College
- CUNY—Laguardia Community College
- CUNY—Medgar Evers College
- CUNY—New York City Community College
- CUNY—York College
- College of Mount St. Vincent
- Columbia University
- Long Island University—Brooklyn
- Manhattan College
- New York Institute of Technology—Main Campus
- New York University
- Pace University
- Pratt Institute
- St. Francis College
- St. Joseph's College—Main Campus
- St. Thomas Aquinas College
- Skidmore College
- SUNY—Stony Brook Main Campus
- SUNY Agricultural and Technical College—Farmingdale
- SUNY Downstate Medical Center
- Suffolk County Community College

Rensselaer Polytechnic Institute
- College of St. Rose
- Hudson Valley Community College
- Russell Sage Junior College—Albany
- Southern Vermont College (Vermont)
- SUNY—Albany

Rochester Institute of Technology
- Bryant-Stratton Business Institute
- Eisenhower College
- Elmira College
- Keuka College
- Monroe Community College
- Nazareth College of Rochester
- Roberts Wesleyan College
- St. John Fisher College
- SUNY—Geneseo
- University of Rochester
- St. Bonaventure University
- Alfred University
- Houghton College
- Jamestown Community College
- SUNY Agricultural and Technical College—Alfred
- University of Pittsburg—Bradford
- St. Johns University
- Academy of Aeronautics
- CUNY—Bernard M. Baruch College
- CUNY—Brooklyn College
- CUNY—City College
- CUNY—Staten Island College
- CUNY—Hunter College
- CUNY—John Jay College of Criminal Justice
- CUNY—Laguardia Community College
- CUNY—New York City Community College
- CUNY—Queens College
- CUNY—Queensborough Community College
- CUNY—York College
- College of Insurance
- Long Island University—C. W. Post Center
- New York Institute of Technology—Main Campus
- New York Institute of Technology—New York City
- New York University
- Pace University
- St. Francis College
- St. Joseph's College—Main Campus
- School of Visual Arts
- Suffolk County Community College
- Wagner College
- St. Lawrence University
- Mater Dei College
- SUNY Agricultural and Technicial College—Canton

Siena College
- Adirondack Community College
- Albany College of Pharmacy
- Albany Business College
- Colgate University
- College of St. Rose
- Fulton Montgomery Community College
- Hudson Valley Community College
- Marist College
- Russell Sage Junior College—Albany
- Schenectady County Community College
- Skidmore College
- SUNY—Albany
- SUNY—New Paltz
- SUNY Agricultural and Technical College—Cobleskill
- SUNY Agricultural and Technical College—Delhi
- SUNY—Empire State College
- Ulster Community College
- Union College

SUNY—Brockport
- Genessee Community College

SUNY—Fredonia
- Jamestown Community College

Syracuse University
- Cayuga County Community College
- Hamilton College
- Herkimer County Community College
- Ladycliff College
- Le Moyne College
- Mohawk Valley Community College
- Onondaga Community College
- SUNY—Morrisville
- SUNY—Buffalo
- SUNY—Oswego
- SUNY College of Environmental Science and Forestry

New York *cont.*
SUNY College of Technology—Utica-Rome
Utica College of Syracuse University
Utica School of Commerce

NORTH CAROLINA
Appalachian State University
Lees McRae College
Campbell University
Central Carolina Technical College
Methodist College
Pembroke State University
Davidson College
Barber-Scotia College
Belmont Abbey College
Catawba College
Central Piedmont Community College
Johnson C. Smith College
Lenoir-Rhyne College
Livingstone College
Mitchell Community College
Pfeiffer College
Queens College
University of North Carolina at Charlotte
Wingate College
Winthrop College
Duke University
North Carolina Central University
University of North Carolina—Chapel Hill
North Carolina State University—Raleigh
Atlantic Christian College
East Carolina University
Methodist College
North Carolina Central University
University of North Carolina—Chapel Hill
St. Augustine's College
Elizabeth City State University
North Carolina Wesleyan College
North Central University
Shaw University
Wake Forest University
High Point College
Winston-Salem State University
Western Carolina University
Brevard College
University of North Carolina—Wilmington

NORTH DAKOTA
University of North Dakota

OHIO
Bowling Green State University
Findlay College
Ohio Northern University
Central State University
Cedarville College
Wilberforce University
Wittenberg University
John Carroll University
Baldwin Wallace College
Case-Western Reserve University
Cleveland State University
Dyke College
Notre Dame College
Ursuline College
Kent State University
Mount Union College
Ohio State University
Capital University
Columbus Technical Institute
Denison University
Devry Institute of Technology
Franklin University
Ohio Dominican College
Ohio Wesleyan University
Otterbein College
Ohio University
Hocking Technical College
Marietta College
Rio Grande College
University of Akron
College of Wooster
The University of Cincinnati
The University of Dayton
Indiana University East (Indiana)
Sinclair Community College
Wright State University
University of Toledo
Xavier University
College of Mt. Joseph on the Ohio
Miami University—Oxford
Northern Kentucky University (Kentucky)
Thomas More College (Kentucky)
Youngstown State University
Westminster College (Pennsylvania)

OKLAHOMA
Cameron University
Western Oklahoma State University
Central State University
Bethany Nazarene College
Langston University
Oklahoma Christian College
East Central University
Southeastern Oklahoma State University
Northwestern Oklahoma State University
Oklahoma Panhandle State University
Oklahoma State University
University of Tulsa
Southwestern Oklahoma State University
University of Oklahoma

OREGON
Oregon State University
Western Oregon State College
Portland State University
Clackamas Community College
Clark College
Concordian College
Marlhurst College
Mount Hood Community College
Oregon Health Science University
Pacific University
Portland Community College
Reed College
University of Portland
Warner Pacific College
University of Oregon
Lane Community College
Northwest Christian College
Oregon Institute of Technology
Southern Oregon State College

PENNSYLVANIA
Bucknell University
Lycoming College
Susquehanna University
University of Pennsylvania—Bloomsburg
University of Pennsylvania—Mansfield
Williamsport Area Community College
Carnegie-Mellon University

La Roche College
St. Vincent College
Clarion University of Pennsylvania
Dickinson College
Harrisburg Area Community College
Lebanon Valley College
Pennsylvania State College—Capital Campus
Drexel University
Beaver College
Bucks County Community College
Hahneman Medical College
Philadelphia College of Osteopathic Medicine
Thomas Jefferson College
Duquesne University
Community College of Allegheny County—Allegheny
Community College of Allegheny County—Boyce
La Roche College
Pennsylvania State University—Beaver
Pennsylvania State University—New Kensington
Point Park College
Robert Morris College
Gannon University
Allegheny College
Alliance College
Edinboro University of Pennsylvania
Mercyhurst College
Mount Aloysius Junior College
Pennsylvania State University—Behrend
Villa Maria College
Gettysburg College
Franklin and Marshall University
Millersville University of Pennsylvania
Mount St. Mary's College (Maryland)
Pennsylvania State University—York
York College of Pennsylvania
Indiana University of Pennsylvania
St. Francis College
Slippery Rock University
Lafayette College
East Stroudsburg University
La Salle College
Beaver College
Community College of Philadelphia
Coombs College of Music
Delaware Valley College of Science and Agriculture
Gwynedd-Mercy College
Hahneman Medical College
Montgomery County Community College
Pennsylvania College of Optometry
Philadelphia College of Bible
Philadelphia College of Textiles and Science
St. Joseph's University
Spring Garden College
Temple University
Lehigh University
Albright College
Allentown College of St. Francis de Sales
Cedar Crest College
Kutztown State College
Lehigh County Community College
Moravian College
Muhlenberg College
Northampton County Area Community College
Pennsylvania State University—Allentown
Pennsylvania State University
Delaware County Community College
Lock Haven University of Pennsylvania
Mansfield State College
Pennsylvania State University—Altoona
Pennsylvania State University—Behrens
Pennsylvania State University—Berks
Pennsylvania State University—Hazelton
Pennsylvania State University—Mont Alto
Pennsylvania State University—Ogontz
Pennsylvania State University—Schuylkill
Pennsylvania State University—Worthington-Scranton
Pennsylvania State University—Delaware
Shippensburg University
Wilson College
Temple University
Bucks County Community College
Chestnut Hill College
Community College of Philadelphia
Delaware Valley College of Science and Agriculture
Drexel University
Montgomery County Community College
Northeastern Christian Junior College
Philadelphia College of Art
Spring Garden College
Trenton State College (New Jersey)
University of Pennsylvania
Glassboro State College (New Jersey)
Philadelphia College of Pharmacy and Science
Rutgers University—Camden (New Jersey)
Villanova University
University of Pittsburgh
Allegheny Community College
Butler County Community College
Carlow College
Chatham College
Community College of Allegheny County—Allegheny
Community College of Beaver County
La Roche College
Pennsylvania State University—McKeesport
Pennsylvania State University—New Kensington
Pittsburgh Theology Seminary
St. Vincent College
University of Pittsburgh—Greensburg
University of Scranton
Baptist Bible College of Pennsylvania
College Misericordia
Keystone Junior College
King's College
Lackawanna Junior College
Luzerne County Community College
Marywood College

Pennsylvania *cont.*
Pennsylvania State University—Wilkes-Barre
Wilkes College
Valley Forge Military Academy and Junior College
Cabrini College
Eastern College
Villanova University
Washington and Jefferson College
California University of Pennsylvania
Waynesburg College
Widener University
Cheney State College
Lincoln University
Pennsylvania State University—Ogontz
Villanova University
West Chester State College

PUERTO RICO
University of Puerto Rico—Mayaguez
American College
Antillian College
Catholic University of Puerto Rico
Inter American University of Puerto Rico—Aguadilla
Inter American University of Puerto Rico—Arecibo
Inter American University of Puerto Rico—Ponce
Inter American University of Puerto Rico—San German
International Institute of the Americas
University of Puerto Rico—Rio Piedras
American College of Puerto Rico
Antillian College
Bayamon Central University
Caguas City College
Carribean University College
Catholic University Metropolitano
Catholic University of Puerto Rico
Colegio Universitario del Turabo
Fundacion Educativa System
Institute of Commercial Design, Puerto Rico Junior College
Instituto Technico Commercial Junior College
Inter American Unversity of Puerto Rico—all campuses
International Institute of the Americas World University
Puerto Rico Junior College
Ramirez College of Business and Technology
San Juan Technological Community College
Universario Polytechnica de Puerto Rico
University of Puerto Rico—all campuses
University of the Sacred Heart

RHODE ISLAND
Providence College
Barrington College
Bridgewater State College (Massachusetts)
Bristol Community College (Massachusetts)
Brown University
Bryant College of Business Administration
Johnson and Wales College
Rhode Island College
Rhode Island Junior College
Roger Williams College—Main Campus
Roger Williams College—Providence
Salve Regina—Newport College
Southwestern Massachusetts University (Massachusetts)
Wheaton College (Massachusetts)
University of Rhode Island

SOUTH CAROLINA
Citadel
Clemson University
Anderson College
Central Wesleyan College
Tri-County Technical College
Furman University
Brevard College (North Carolina)
Greenville Technical College
North Greenville College
Presbyterian College
Lander College
Newberry College
South Carolina State College
Claflin College
Denmark Technical College
Voorhees College
University of South Carolina
Benedict College
Francis Marion College
Morris College
University of South Carolina—Coastal Carolina
Wofford College
Converse College
Erskine College and Seminary
Gardner-Webb College (North Carolina)
Spartanburg Methodist College
University of South Carolina—Spartanburg

SOUTH DAKOTA
South Dakota School of Mines and Technology
Black Hills State College
National College
South Dakota State University
Dakota State College
Northern State College
University of South Dakota
Mount Marty College

TENNESSEE
Austin Peay State University
Carson-Newman College
Lincoln Memorial University
East Tennessee State University
King College
Milligan College
Tusculum College
Memphis State University
Christian Brothers College
Le Moyne–Owen College
Rhodes College
Rust College (Mississippi)
Shelby State Community College
Middle Tennessee State University
Tennessee Technological University
University of Tennessee—Chattanooga
Bryan College
Tennessee Temple University

University of Tennessee—Knoxville
- Knoxville College

University of Tennessee—Martin

Vanderbilt University
- Aquinas Junior College
- Belmont College
- David Lipscomb College
- Fisk University
- Tennessee State University
- Trevecca Nazarene College
- Volunteer State Community College

TEXAS

Bishop College
- Dallas Baptist College
- Jarvis Christian College
- University of Texas at Dallas

Hardin-Simmons College
- Abilene Christian University
- Cisco Junior College
- McMurray College

Midwestern State University

Pan American University
- Texas Southmost College

Prairie View A & M University
- St. Mary's University
- Incarnate Word College
- Our Lady of the Lake University
- San Antonio College

Sam Houston State University

Southern University A & M College

Stephen F. Austin State University

Texas A & I University
- Corpus Christi State University
- Del Mar College

Texas A & M University
- Tarleton State University

Texas Christian University
- North Lake College
- Southwestern Baptist Seminary
- Tarrant County Junior College Northeast
- Tarrant County Junior College Northwest
- Tarrant County Junior College South
- Texas Wesleyan College
- University of Dallas

Texas Technological University
- Lubbock Christian College

Trinity University
- Incarnate Word College
- St. Phillips College
- San Antonio College
- University of Texas—San Antonio

University of Houston
- Alvin Community College
- Houston Baptist University
- Houston Community College
- Rice University
- San Jacinto College
- Texas Southern University
- Texas Women's University
- University of Houston—Clear Lake City
- University of Houston—Downtown College
- University of St. Thomas
- University of Texas—Health Sciences Center

University of Texas at Arlington
- Brookhaven College
- Eastfield College
- El Centro College
- Mountain View College
- Navarro College
- North Texas State University
- Richland College
- Southern Methodist University
- Texas Women's University

University of Texas at Austin
- American Technology University
- Austin Community College
- Concordia Lutheran College
- Huston-Tillotson College
- St. Edwards University
- Southwest Texas State University

University of Texas at El Paso

University of Texas—San Antonio
- San Antonio College
- University of Texas Health Sciences Center

West Texas State University

UTAH

Brigham Young University
- Utah Technical College

University of Utah
- Westminster College

Utah State University

Weber State College

VERMONT

Norwich University

Dartmouth College (New Hampshire)

Norwich University—Vermont College

University of Vermont
- Castleton State College
- Champlain College
- Johnson State College
- Middlebury College
- St. Michaels College
- Southern Vermont College
- Trinity College

VIRGINIA

College of William and Mary
- Christopher Newport College

Hampton Institute
- Thomas Nelson Community College

James Madison University
- Blue Ridge Comunity College
- Mary Baldwin College

Old Dominion University
- Tidewater Commmunity College

University of Richmond
- Hampton-Sidney College
- J. Sargeant Reynolds Community College
- Longwood College
- Randolph-Macon Women's College
- Virginia Commonwealth University

Virginia Military Institute

Virginia Polytechnic Institute and State University
- Ferrum College
- Radford University

Virgina State University
- John Tyler Community College
- Saint Paul's College
- Virginia Union University

Norfolk State University

Washington and Lee University
- Central Virginia Community College
- Liberty Baptist College
- Lynchburg College
- Randolph-Macon Woman's College
- Sweet Briar College

Virginia *con't.*
University of Virginia
Piedmont Valley Community College

WASHINGTON
Eastern Washington University
Central Washington University
Gonzaga University
North Idaho College (Idaho)
Spokane Community College
Spokane Falls Community College
Whitworth College
Seattle University
Bellevue Community College
City College
Fort Steilacoom Community College
Green River Community College
Highline Community College
Olympic College
Pacific Lutheran University
St. Martin's College
Seattle Community College—Central
Seattle Community College—South
Seattle Pacific University
Tacoma Community College
University of Puget Sound
University of Washington
Bellevue Community College
Everett Community College
Highline Community College
North Seattle Community College
Pacific Lutheran University
Seattle Pacific University
Shoreline Community College
Washington State University
Eastern Oregon State College (Oregon)

WEST VIRGINIA
Marshall University
West Virginia State College
University of Charleston
West Virginia Institute of Technology
West Virginia University
Fairmont State College
Frostburg State College (Maryland)

WISCONSIN
Marquette University
Alverno College
Carthage College
Lakeland College
Milwaukee School of Engineering
Mount Mary College
University of Wisconsin in Parkside
Ripon College
Marian College of Fond du Lac
University of Wisconsin—Fond du Lac
Saint Norbert College
University of Wisconsin—Green Bay
University of Wisconsin—Madison
University of Wisconsin—Milwaukee
Cardinal Stritch College
Milwaukee School of Engineering
University of Wisconsin—Fond du Lac
University of Wisconsin—Waukesha County
University of Wisconsin—La Crosse
Viterbo College
University of Wisconsin—Platteville
Clarke College (Iowa)
Loras College (Iowa)
University of Dubuque (Iowa)
University of Wisconsin—Stevens Point
University of Wisconsin—Whitewater

WYOMING
University of Wyoming

Navy and Marine Corps

ALABAMA
Auburn University
Pima Community College

ARIZONA
University of Arizona

CALIFORNIA
University of California—Berkeley
California Maritime Academy—Vallejo
California State University—Hayward
California State University—Sacramento
California State University—San Jose
San Francisco State University
Stanford University
University of California—Davis
University of San Francisco
University of Santa Clara
University of California—Los Angeles
California State University—Fullerton
California State University—Long Beach
California State University—Los Angeles
California State University—Northridge
Loyola Marymount University
Northrop University
Occidental College
Pepperdine University
University of California—Irvine
University of San Diego, San Diego State University
University of California—San Diego
University of Southern California
California Institute of Technology
California State Polytechnic University
Claremont McKenna College
Harvey Mudd College

COLORADO
University of Colorado

DISTRICT OF COLUMBIA
George Washington University
American University

Catholic University of America
Georgetown University
Howard University
University of the District of Columbia

FLORIDA
Florida A & M University
Florida State University
Tallahassee Community College
Jacksonville University
Florida Junior College—Jacksonville
University of North Florida
University of Florida

GEORGIA
Georgia Institute of Technology
Agnes Scott College
Clark College
Georgia State University
Kennesaw College
Morehouse College
Morris Brown College
Oglethorpe University
Southern Technical Institute
Spelman College
Savannah State College
Armstrong State College

IDAHO
University of Idaho
Washington State University

ILLINOIS
Illinois Institute of Technology
Elmhurst College
Prairie State College
Purdue University—Calumet (Indiana)
University of Chicago
University of Illinois—Chicago
Wilbur Wright City College
Northwestern University
Loyola University
North Park College
University of Illinois
Parkland College

INDIANA
Purdue University
University of Notre Dame
Bethel College
Indiana University—South Bend

IOWA
Iowa State University

KANSAS
University of Kansas

LOUISIANA
Southern University A & M
Louisiana State University
Tulane University

MAINE
Maine Maritime Academy (Navy only)
University of Maine—Orono

MASSACHUSETTS
Boston University
Northeastern University
College of the Holy Cross
Anna Maria College
Assumption College
Central New England College
Clark University
Worcester Polytechnic Institute
Worcester State College
Massachusetts Institute of Technology
Harvard University
Tufts University
Wellesley College

MICHIGAN
University of Michigan
Eastern Michigan Universiity—Ypsilanti

MINNESOTA
University of Minnesota
Augsburg College
College of St. Thomas
Macalester College

MISSISSIPPI
University of Mississippi

MISSOURI
University of Missouri
Columbia College

NEBRASKA
University of Nebraska
Nebraska Wesleyan University

NEW MEXICO
University of New Mexico

NEW YORK
Cornell University
Ithaca College
SUNY—Cortland
SUNY—Maritime College (Navy only)
Fordham University
Iona College
Manhattan College
Rensselaer Polytechnic Institute
Russell Sage College
Siena Collge
Skidmore College
SUNY—Albany
Union College
University of Rochester
Monroe Community College
Nazareth College
Rochester Institute of Technology
St. John Fisher College
SUNY—Brockport
SUNY—Genesee

NORTH CAROLINA
Duke University
North Carolina Central University
University of North Carolina
North Carolina State University

OHIO
Miami University
Ohio State University
Capital University
Ohio Dominican College
Otterbein University

OKLAHOMA
University of Oklahoma

OREGON
Oregon State University
Linn Benton Community College

PENNSYLVANIA
Pennsylvania State University—University Park
University of Pennsylvania—Philadelphia
Drexel University
La Salle College
St. Joseph's University
Temple University
Villanova University
Temple University

SOUTH CAROLINA
Citadel
University of South Carolina

TENNESSEE
Memphis State University
Vanderbilt University
Belmont College
David Lipscomb College
Fisk University
Tennessee State University
Trevecca Nazarene College

TEXAS
Prairie View A & M University
Rice University
Houston Baptist University
Texas Southern University
University of Houston
University of St. Thomas
Texas A & M University
Texas Technological University
University of Texas

UTAH
University of Utah
Weber State College
Westminster College

VERMONT
Norwich University

VIRGINIA
Hampton Institute, Norfolk State University, Old Dominion University
University of Virginia
Piedmont Community College
Virginia Military Institute
Virginia Polytechnic Institute and State University

WASHINGTON
University of Washington
Seattle Pacific University
Seattle University

WISCONSIN
Marquette University
Milwaukee School of Engineering
Mt. Mary College
University of Wisconsin—Milwaukee
University of Wisconsin

Air Force

ALABAMA
Alabama State University
Auburn University at Montgomery
Huntingdon College
Troy State University in Montgomery
Auburn University
Samford University
Birmingham Southern College
Jefferson State Junior College
Lawson State Community College
Miles College
University of Alabama—Birmingham
University of Montevallo
Troy State University
Tuskegee Institute
University of Alabama—University

ARIZONA
Arizona State University
Glendale Community College
Grand Canyon College
Mesa Community College
Phoenix College
Scottsdale Community College
South Mountain Community College
Embry-Riddle Aeronautical at Prescott
Yavapai College
Northern Arizona University
University of Arizona
Pima Community College

ARKANSAS
University of Arkansas

CALIFORNIA
California State University—Fresno
College of the Sequoias
Fresno City College
Kings River College
Merced College
West Coast Bible College
West Hills College
California State University—Long Beach
California State College—San Bernardino
California State Polytechnic University
California State University—Dominguez Hills
California State University—Fullerton
California State University—Los Angeles
Chaffey College
Citrus College
Cypress College
El Camino College
Fullerton College
Golden West College
Long Beach City College
Los Angeles Harbor College
Mount San Jacinto College
Orange Coast College
Pasadena City College
Rio Hondo College
Riverside City College
Saddleback Community College
San Bernardino Valley College
Santa Ana College
University of California—Riverside
California State University—Sacramento
American River College
Butte Junior College
Consumnes River College
Sacramento City College
Sierra College
Solano Junior College
University of California—Davis
University of the Pacific
Yuba College

Loyola Marymount University
Antelope Valley College
California State College—San Bernardino
California State Polytechnic University
California State University—Dominguez Hills
California State University—Fullerton
California State University—Long Beach
California State University—Los Angeles
California State University—Northridge
Chapman College
Cypress College
East Los Angeles College
El Camino College
Fullerton College

Golden West College
Los Angeles City College
Los Angeles Harbor College
Los Angeles Pierce College
Los Angeles Southwest College
Los Angeles Trade and Technical College
Los Angeles Valley College
Marymount Palos Verdes College
Moorpark College
Mount St. Mary's College
Mount San Jacinto College
Northrop University
Orange Coast College
Pasadena City College
Pepperdine University
Rio Hondo College
Riverside City College
Saddleback Community College
San Bernardino Valley College
Santa Ana College
Santa Monica College
University of California—Irvine
University of California—Riverside
University of Redlands
Victor Valley College
West Los Angeles College
Westmont College

San Diego State University
Cayamaca Community College
Grossmont Community College
National University
Palomar College
Point Loma College
San Diego Community College—City College
San Diego Community College—Evening College
San Diego Community College—Mesa College
San Diego Community College—Miramar College
Southwestern College
University of California—San Diego
University of San Diego

San Francisco State University
Cogswell College
Dominican College of San Rafael
Golden Gate University
University of California—Hastings College of Law
University of California—San Francisco
University of California—San Francisco—Medical Center

San Jose State University
Cabrillo College
De Anza College
Evergreen Valley College
Foothill College
Mission College
Ohlone College
San Jose City College
Stanford University
University of Santa Clara
West Valley College
West Valley Joint Community College District

University of California—Berkeley
California State University—Hayward
Chabot College
City College of San Francisco
College of Alameda
Contra Costa College
Diablo Valley College
Graduate Theological Union
Holy Names College
Laney College
Los Medanos College
Merritt College
Mills College
Ohlone College
St. Mary's College
Sonoma State College

University of California—Los Angeles
California Lutheran College
California State College—San Bernardino
California State Polytechnic University
California State University—Dominguez Hills
California State University—Fullerton
California State University—Long Beach
California State University—Los Angeles
California State University—Northridge
Los Angeles Mission College
Mount St. Mary's College
Northrop University
Santa Monica College
University of California—Irvine
University of California—Riverside
University of California—Santa Barbara
University of La Verne

University of Southern California
Biola College
California Institute of Technology
California Lutheran College
California State College—San Bernardino
California State Polytechnic University
California State University—Dominguez Hills
California State University—Fullerton
California State University—Los Angeles
California State University—Northridge
Chaffey College
Chapman College
Citrus College
Claremont Men's College
Compton Community College
Cypress College
East Los Angeles College
El Camino College
Fullerton College
Glendale Community College
Golden West College
Harvey Mudd College
Long Beach City College
Los Angeles Harbor College
Los Angeles Pierce College
Los Angeles Southwest College
Los Angeles Trade and Technical College
Los Angeles Valley College
Moorpark College
Mount San Antonio College
Northrop University
Occidental College
Orange Coast College
Pasadena City College

California *cont.*
Pepperdine University
Pomono College
Rio Hondo College
San Bernardino Valley College
University of California—Irvine
University of California—Riverside
Ventura College
West Los Angeles College
Whittier College

COLORADO
Colorado State University
University of Northern Colorado
Aims Community College
University of Colorado
Arapahoe Community College
Colorado School of Mines
Metropolitan State College
Regis College
University of Colorado—Denver
University of Colorado Health Sciences Center
University of Denver

CONNECTICUT
University of Connecticut
Central Connecticut State College
Eastern Connecticut State College
Southern Connecticut State College
Trinity College
University of Hartford
Western Connecticut State College

DELAWARE
University of Delaware
Wilmington College

DISTRICT OF COLUMBIA
Howard University
American University
Catholic University of America
Georgetown University
George Washington University
Trinity College
University of the District of Columbia

FLORIDA
Embry-Riddle Aeronautical University
Bethune-Cookman College
Daytona Beach Community College
University of Central Florida
Florida State University
Florida A & M University
Tallahassee Community College
University of Central Florida
Brevard Community College
Florida Southern College
Lake-Sumter Community College
Rollins College
Seminole Community College
Valencia Community College
University of Florida
Santa Fe Community College
University of Miami
Barry College
Bicayne College
Florida International University
Florida Memorial College
Miami-Dade Community College
University of South Florida
Hillsborough Community College
Pasco-Hernando Community College
Polk Community College

GEORGIA
Georgia Institute of Technology
Agnes Scott College
Clark College
Georgia State University
Morehouse College
Morris Brown College
Southern Technical Institute
Spelman College
University of Georgia
Medical College of Georgia
Valdosta State College

HAWAII
University of Hawaii—Manoa
Chaminade University of Honolulu
Hawaii Pacific College
Honolulu Community College
Kapiolani Community College
Leeward Community College
West Oahu College
Windward Community College

ILLINOIS
Illinois Institute of Technology
Chicago State University
Elmhurst College
Governors State University
John Marshall Law School
Kennedy-King College
Lewis University
Loop College
Malcolm X College
North Central College
Northeastern Illinois University
Northern Illinois University
Northwestern University
Olive-Harvey College
Richard J. Daley College
Rush University
St. Xavier College
Triton College
Truman College
University of Illinois—Chicago Circle
University of Illinois Medical Center—Chicago
Wright College
Parks College of St. Louis University
Harris Stowe State College (Missouri)
St. Louis Community College (Missouri)
St. Louis University (Missouri)
University of Missouri (Missouri)
Washington University (Missouri)
Southern Illinois University—Edwardsville
Belleville Area College
Lewis and Clark Community College
McKendree College
University of Illinois—Urbana
Parkland College

INDIANA
Indiana University
Butler University
DePauw University
Indiana State University

Indiana University, Purdue University—Indianapolis
Marian College
Rose-Hulman Institute of Technology
Purdue University
University of Notre Dame
Holy Cross Junior College
Indiana University—South Bend
St. Mary's College

IOWA
Iowa State University
Drake University
University of Iowa

KANSAS
Kansas State University
Mid-America Nazarene College
University of Kansas
Washburn University

KENTUCKY
University of Kentucky
Eastern Kentucky University
Georgetown College
Kentucky State University
Midway College
Transylvania University
University of Louisville
Bellarmine College
Indiana University Southeast
Louisville Presbyterian Theological Seminary
Southern Baptist Theological Seminary
Spalding College

LOUISIANA
Grambling State University
Louisiana State University and A & M College
Southern University A & M College
Louisiana Technical University
University of New Orleans
Dillard University
Louisiana State University School of Nursing
Loyola University in New Orleans
Our Lady of Holy Cross College
Southern University in New Orleans
Tulane University
Xavier University of Louisiana
University of Southwestern Louisiana

MAINE
University of Maine
Husson College

MARYLAND
University of Maryland
Anne Arundel Community College
Bowie State College
George Mason University (Virginia)
Johns Hopkins University
Loyola College
Shepherd College (West Virginia)
Towson State University
Western Maryland College

MASSACHUSETTS
Boston University
Northeastern University
College of the Holy Cross
Anna Maria College
Assumption College
Becker Junior College—Leicester
Becker Junior College—Worcester
Central New England College
Clark University
Quinsigamond Community College
Worcester Polytechnic Institute
Worcester State College
Massachusetts Institute of Technology
Harvard University
Tufts University
Wellesley College
University of Lowell
Bentley College
Daniel Webster College (New Hampshire)
Endicott College
Gordon College
Middlesex Community College
New England College (New Hampshire)
New Hampshire College (New Hampshire)
Northern Essex Community College
North Shore Community College
Notre Dame College (New Hampshire)
River College (New Hampshire)
St. Anselm College (New Hampshire)
Salem State College
University of Massachusetts—Amherst
Amherst College
Mount Holyoke College
Smith College
Western New England College

MICHIGAN
Michigan State University
Lansing Community College
Michigan Technological University
Suomi College
University of Michigan
Concordia College
Eastern Michigan University
Lawrence Institute of Technology
University of Michigan—Dearborn
Wayne State University

MINNESOTA
College of St. Thomas
Anoka-Ramsey Community College
Augsburg College
Bethel College
College of St. Catherine
Hamline University
Inver Hills Community College
Lakewood Community College
Macalester College
Normandale Community College
North Hennepin Community College
William Mitchell College of Law
University of Minnesota—Duluth
College of St. Scholastica
University of Minnesota—Minneapolis

MISSISSIPPI
Mississippi State University—Mississippi State
Mississippi University for Women
Mississippi Valley State University
Delta State University

Mississippi *cont.*
University of Mississippi—University
University of Southern Mississippi
William Carey College

MISSOURI
Southeast Missouri State University
University of Missouri—Columbia
Columbia College
Stephens College
William Woods College
University of Missouri—Rolla

MONTANA
Montana State University

NEBRASKA
University of Nebraska
Concordia Teachers College
Nebraska Wesleyan University
University of Nebraska—Omaha
Bellevue College
College of St. Mary
Creighton University
Iowa Western Community College
University of Nebraska Medical Center

NEW HAMPSHIRE
University of New Hampshire
Nathaniel Hawthorne College
New England College
Plymouth State
St. Anselm's College
University of Southern Maine (Maine)

NEW JERSEY
New Jersey Institute of Technology
Essex County Community College
Fairleigh Dickinson University—Teaneck
Jersey City State College
Kean College of New Jersey
Montclair State College
Rutgers University—Newark
St. Peter's College
Seton Hall University
Stevens Insitute of Technology
William Paterson College of New Jersey
Rutgers, the State University
Brookdale Community College
Mercer County Community College
Middlesex County College
Monmouth College
Princeton University
Rider College
Rutgers University—Camden
Somerset County College
Trenton State College
Union College
Wagner College (New York)

NEW MEXICO
New Mexico State University
University of Texas—El Paso (Texas)
University of New Mexico
University of Albuquerque

NEW YORK
Clarkson University
St. Lawrence University
SUNY—Potsdam
SUNY Agriculture and Technical College—Canton
Cornell University
Ithaca College
SUNY—Cortland
Tompkins Cortland Community College
Wells College
Manhattan College
Academy of Aeronautics
Adelphi University
College of Mount St. Vincent
Columbia University
Dowling College
Elizabeth Seton College
Long Island University—Brooklyn
Long Island University—C. W. Post Center
Mercy College
Molloy College
Nassau Community College
New York Institute of Technology
Pace University
Polytechnic Institute of New York
St. Francis College
St. Joseph's College—Brentwood
St. Thomas Aquinas College
Southampton Community College
Suffolk Community College
SUNY—Old Westbury
SUNY—'Stony Brook
SUNY Agricultural and Technical College—Farmingdale
Rensselaer Polytechnic Institute
Albany College of Pharmacy
College of St. Rose
Fulton-Montgomery Community College
Hudson Valley Community College
Maria College
Russell Sage College
Schenectady County Community College
Siena College
Skidmore College
SUNY—Albany
SUNY—Empire State College
Union College
Syracuse University
Le Moyne College
New School for Social Research
Onondaga Community College
SUNY College of Environmental Science and Forestry
Utica College of Syracuse University

NORTH CAROLINA
Duke University
North Carolina Central University
East Carolina University
Pitt Community College
Fayetteville State University
Pembroke State University
North Carolina Agricultural and Technical State University
Bennett College
Greensboro College
Guilford College

High Point College
University of North Carolina—Greensboro
North Carolina State University—Raleigh
Meredith College
Peace College
St. Augustine's College
St. Mary's College
Shaw University
University of North Carolina—Chapel Hill
University of North Carolina—Charlotte
Barber-Scotia College
Belmont Abbey College
Central Piedmont Community College
Davidson College
Gaston College
Johnson C. Smith University
Sacred Heart College
Queens College
Wingate College
Winthrop College

NORTH DAKOTA

North Dakota State University of Agriculture and Applied Science
Concordia College
Moorhead State University

OHIO

Bowling Green State University
Ashland College
Defiance College
Findlay College
Heidelberg College
Ohio Northern University
University of Toledo
Kent State University
Miami University—Oxford
Miami University—Hamilton
Miami University—Middletown
Ohio State University
Capital University
Franklin University
Ohio Dominican College
Ohio Institute of Technology
Ohio Wesleyan University
Otterbein College
Ohio University
University of Akron
University of Cincinnati
Cincinnati Technical College
College of Mount St. Joseph
Edgecliff College
Northern Kentucky University
Thomas More College
Xavier University
Wright State University
Antioch College
Cedarville College
Central State University
Clark Technical College
Edison State Community College
Sinclair Community College
Southern State Community College
University of Dayton
Urbana College
Wilberforce University
Wilmington College
Wittenberg University

OKLAHOMA

Oklahoma State University
University of Oklahoma
Oklahoma Christian College
Oklahoma City University
Oscar Rose Junior College
St. Gregory's College

OREGON

Oregon State University
Linn-Benton Community College
University of Oregon
Western Oregon State College
University of Portland
Clackamas Community College
Clark College (Washington)
Concordia College
Mount Hood Community College
Oregon Health Sciences Center
Portland Community College
Warner-Pacific College
Willamette University

PENNSYLVANIA

Carnegie-Mellon University
Grove City College
Slippery Rock State College
Lehigh University
Allentown College of St. Francis de Sales
Cedar Crest College
East Stroudsburg State College
Kutztown State College
Lafayette College
Moravian College
Muhlenberg College
Northampton County Area Community College
Pennsylvania State University—Allentown
Pennsylvania State University—Berks
St. Joseph's University
Drexel University
Eastern College
La Salle College
Rutgers University—Camden
Temple University
Thomas Jefferson University
University of Pennsylvania—Philadelphia
Villanova University
West Chester State College
Widener University
Pennsyvlania State University—University Park
University of Pittsburgh
Carlow College
Chatham College
Community College of Allegheny County—Boyce County
Community College of Allegheny County—North Campus
Community College of Allegheny County—Pittsburgh
Community College of Allegheny County—West Mifflin
Duquesne University
La Roche College
Point Park College
Robert Morris College
St. Vincent College
Wilkes College
Bloomsburg State College
College Misericordia
Keystone Junior College
King's College
Lackawanna Junior College
Luzerne County Community College
Marywood College
Pennsylvania State University—Hazelton

Pennsylvania *cont.*
Pennsylvania State University—Wilkes-Barre
Pennsylvania State University—Worthington-Scranton
University of Scranton

PUERTO RICO
University of Puerto Rico—Rio Piedras
Bayamon Central University
Bayamon Regional College
Inter American University of Puerto Rico—Hato Rey
Sacred Heart University
University of Puerto Rico—Bayamon Technical University College
University of Puerto Rico—Cardina Regional College
University of Puerto Rico—Cayey University College
University of Puerto Rico—Humacao University College
World University
University of Puerto Rico—Mayaguez
Inter American University of Puerto Rico—San German
University of Puerto Rico—Aguadilla Regional College

SOUTH CAROLINA
Baptist College at Charleston
College of Charleston
Medical University of South Carolina
South Carolina State College
Citadel
Clemson University
Anderson College
Central Wesleyan College
Greenville Technical College
Tri County Technical College
University of South Carolina
Benedict College

SOUTH DAKOTA
South Dakota State University

TENNESSEE
Memphis State University
Christian Brothers College
LeMoyne-Owen College
Shelby State Community College
Southwestern at Memphis
University of Tennessee Medical School
Tennessee State University
Aquinas Junior College
Belmont College
David Lipscomb College
Fisk University
Meharry Medical College
Middle Tennessee State University
Trevecca Nazarene College
Vanderbilt University
Volunteer State Community College
Western Kentucky University (Kentucky)
University of Tennessee
Knoxville College

TEXAS
Angelo State University
Baylor University
McLennan Community College
Paul Quinn College
University of Mary Hardin-Baylor
East Texas State University
North Texas State University
Southern Methodist University
Texas Woman's University
University of Dallas
University of Texas—Dallas
Southwest Texas State University
Texas Lutheran College
University of Texas—San Antonio
Texas Christian University
Baylor School of Nursing
Tarrant County Junior College
Texas Wesleyan College
University of Texas—Arlington
Texas Technological University
Lubbock Christian College
University of Texas—Austin
Austin Community College
Concordia Lutheran College

UTAH
Brigham Young University
Utah Technical College
University of Utah
Weber State College
Westminster College
Utah State University

VERMONT
Norwich University
St. Michael's College
Champlain College
Lyndon State College
Trinity College
University of Vermont

VIRGINIA
Virginia Military Institute
Virginia Polytechnic Institute
University of Virginia
Piedmont Virginia Community College

WASHINGTON
Central Washington University
University of Puget Sound
Fort Steilacoom Community College
Pacific Lutheran University
St. Martin's College
Southern Illinois University
Tacoma Community College
University of Washington
Bellevue Community College
Everett Community College
Edmonds Community College
Green River Community College
Highline Community College
North Seattle Community College

Seattle Community College
Seattle University
Shoreline Community College
Washington State University
University of Idaho (Idaho)

WEST VIRGINIA
West Virginia University
Fairmont State College

WISCONSIN
University of Wisconsin—Madison
University of Wisconsin—Superior

WYOMING
University of Wyoming

Index